G. BROWN.

£25.

E74.

Nuclear magnetic resonance and its applications to living systems

To Ro

Nuclear magnetic resonance and its applications to living systems

BY

DAVID G. GADIAN
Department of Biochemistry, University of Oxford

CLARENDON PRESS · OXFORD
OXFORD UNIVERSITY PRESS · NEW YORK

Oxford University Press, Walton Street, Oxford OX2 6DP

London New York Toronto
Delhi Bombay Calcutta Madras Karachi
Kuala Lumpur Singapore Hong Kong Tokyo
Nairobi Dar es Salaam Cape Town
Melbourne Auckland

and associated companies in
Beirut Berlin Ibadan Mexico City Nicosia

Oxford is a trade mark of Oxford University Press

Published in the United States
by Oxford University Press, New York

British Library Cataloguing in Publication Data
Gadian, David G.
Nuclear magnetic resonance and its applications to living systems.
1. Nuclear magnetic resonance spectroscopy
2. Biological physics — Technique
I. Title
574'.028 QH324.9
ISBN 0-19-854627-0

Printed in Great Britain by
The Alden Press Ltd.
Oxford, London and Northampton

Preface

In recent years, nuclear magnetic resonance (n.m.r.) has emerged as a novel method of studying the metabolism and anatomical structure of intact biological systems. The method has the rather unusual and attractive features that it is totally non-destructive and non-invasive, and for these reasons it has interesting possibilities in the fields of biology and medicine. This book introduces the technique of n.m.r., and describes how it can be used to study the metabolism of a wide variety of living systems, ranging from bacteria to human beings. In addition, an explanation is given of the way in which n.m.r. can provide anatomical images of human beings, analogous to those obtained by X-ray computed tomography.

This book is written primarily for biologists, many of whom will be unfamiliar with n.m.r. Therefore I have attempted to introduce the theory and practice of the technique in a reasonably simple and yet informative manner, bearing in mind throughout that the reader may have relatively little background in the physical sciences. The question was how to strike the right balance between the n.m.r. and the biology, for I wished to preserve a strong biological theme in the book. After some thought and advice, I decided to adopt the following approach. In the first chapter, I give a brief introduction to n.m.r. and its biological applications, and summarize the material that is covered in much greater depth in the subsequent chapters. The next three chapters have a strong emphasis on the biology. First, I describe in general terms the type of information that the biologist can obtain from n.m.r. studies of living systems. Then I discuss some specific applications to cells, tissues, whole animals, and human beings, and I have divided the material in such a way that Chapter 3 should appeal rather more to biochemists and Chapter 4 to clinicians. I have not attempted to cover the literature comprehensively, or to acknowledge all of the contributions to this field of research, for if I did the book would read far too much like a review. In view of the rapid developments that are taking place, I (and the publishers) have made every effort to ensure that the material discussed in these chapters is fully up-to-date. Certainly, there are many references to work published in 1981.

The latter half of the book is concerned with n.m.r. theory, instrumentation, and operating procedures. It could be argued that these topics are adequately discussed elsewhere, and that I have included these chapters merely for the sake of completeness. However, many biologists entering the field of n.m.r. are daunted by the apparent complexity of the technique and of the books that are devoted to it. Therefore I feel that there is a pressing need for a coherent account of the theory and practice of n.m.r. that is written at a level suitable for biologists. In the last four chapters, I have attempted to present such an account. Chapters 5 and 6 describe the basic theory of the technique, and discuss the meaning and measurement of the various parameters that characterize n.m.r. spectra. Emphasis is placed on the role of the Fourier transform in n.m.r., and indeed Fourier transform n.m.r. is the only mode of detection that is discussed in any detail.

The final two chapters describe n.m.r. instrumentation, and are intended primarily for research workers who wish to use the technique. All too often it is found that users of n.m.r. spectrometers have surprisingly little appreciation of how they work, and are therefore unable to get the most out of the instruments. These chapters should help to overcome the problem, for they outline the principles and logic of an n.m.r. spectrometer at a level that should be suitable for a non-technical audience. The last chapter should be particularly useful for anyone who wishes to set up n.m.r. studies of living systems, for it describes in some detail the remarkable improvements in spectrometer performance and versatility that can be achieved by making relatively minor modifications to the design of the n.m.r. probes that contain the samples.

Among chemists and biochemists, n.m.r. is probably most widely known for its ability to provide detailed information about the structures of molecules in solution. This type of application has been extensively discussed in several books, and is not described here. I have also excluded n.m.r. studies of membranes, partly because many of these studies have been performed on preparations that cannot be classed as living systems, but also because an understanding of much of this work would require a more extensive knowledge of n.m.r. than can be obtained from a book of this type.

This is an excellent opportunity to acknowledge my debt to all the many colleagues and friends with whom I have worked over the last ten years, who have introduced me to n.m.r. and its biological applications, who have taught me much that I know about the subject, and who in general have provided a most stimulating and enjoyable atmosphere in which to work. These colleagues are far too numerous for mention to be made of them all, but in particular I owe a great deal to Dr G.K. Radda, Sir Rex Richards, Dr D.I. Hoult, Mr P. Styles, Dr Joan Dawson, and Professor D.R. Wilkie. I am also most grateful to everyone who made valuable criticisms and comments about the manuscript.

Finally, I would like to acknowledge all those who granted permission to reproduce diagrams; details are given in the figure captions and references. I am particularly grateful to Drs Brown and Campbell, and to Drs Hawkes, Holland, and Moore for providing figures that had not been published, and also to Dr Morehead for providing the original drawing for Plate 1(b).

Oxford   D.G.G.

May 1981

Contents

List of plates

1

Introduction

1.1. A non-destructive method of studying living systems

We now have a fairly detailed understanding of the mechanisms that control the activity of some individual enzymes. Biochemists are therefore becoming increasingly interested in studying the organization and control of integrated metabolic pathways, and in relating their observations both to the activities of the constituent enzymes of the pathways and also to the physiological state of the system under investigation. There are two distinct approaches to the study of integrated enzyme systems, one of which is to study the pathway *in vitro* under conditions that resemble as closely as possible those that are expected *in vivo*. This approach has the virtues of simplicity and of enabling the experimenter to control at will the conditions under which the pathway is studied. However, any conclusions that are reached must eventually be confirmed by observations of the *in vivo* system. The second approach, which is to study the metabolic pathway *in vivo*, is direct and enables the observed metabolic activity to be related to the physiological state of the system. However, the measurement of changes in metabolite levels within living systems is not an easy matter.

Metabolite levels in living tissues and organs are commonly measured using the technique known as 'freeze clamping'. This method involves rapid freezing of the tissue, extraction of the soluble components, and subsequent analysis for the required compounds. The usefulness of the method cannot be understated, but it does have several limitations. For example, by destroying the integrity of the tissue information is lost about the environment of the molecules within the cell. Furthermore, great care has to be taken to ensure that an insignificant amount of metabolism takes place during the interval between freezing the tissue and analysis. In addition, the time course of metabolic processes cannot be followed within a single preparation because of the destructive nature of the technique.

The emergence of n.m.r. as a non-destructive method of studying the metabolism of cells and tissues is therefore to be welcomed by the biochemist. In this book, we describe how n.m.r. can be used to study the metabolism of a wide range of living systems. We discuss the results that have been obtained, the scope of the method, and its limitations which are admittedly fairly severe. We stress that n.m.r. provides information that is often complementary to that obtained by more traditional techniques, and therefore that it should be used, not to the exclusion of other analytical techniques, but rather in conjunction with any other methods that may be of value.

N.m.r. could also be of considerable value to the clinician, as a result of two parallel, and perhaps complementary developments that have been taking place in recent years. Firstly, it has become apparent that metabolic studies using n.m.r. need not be restricted to cell, tissue, and organ preparations; whole animals and human beings can also be investigated. Thus the technique now provides a totally non-invasive method of studying the metabolic state of human tissues. The second development involves the use of n.m.r. as a method of obtaining images of human beings, analogous in many ways to those obtained with X-ray computed tomography (CT) scanners. At the time of writing, clinical trials using these two n.m.r. approaches have only just begun. It is therefore too soon to predict the eventual value of n.m.r. to the clinician, but on the basis of preliminary results that are described in Chapter 4, the outlook seems most promising.

The purpose of this introductory chapter is to provide a brief overview of the material that is to be covered in the subsequent chapters. We begin by giving an introduction to the field of spectroscopy in general, and to n.m.r. in particular. We then discuss briefly the type of information that the biologist can obtain using n.m.r., and illustrate the scope of n.m.r. studies of living systems by describing a few examples from recent literature in the field.

1.2. An introduction to spectroscopy

In 1666 Isaac Newton obtained 'a triangular glass prism to try therewith the celebrated phenomena of colours'. He showed that the prism separated white light into what he called a 'spectrum' of the colours, and his experiments marked the beginnings of the branch of science that we term spectroscopy.

Spectroscopy deals with the interaction of electromagnetic radiation with matter. The radiation that is absorbed or emitted by the sample is detected in some way, and the output of the detector displayed as a function of the frequency or wavelength of the radiation is termed a spectrum. The spectrum can be interpreted in terms of some aspect of the atomic or molecular structure of the material under investigation.

The electromagnetic theory developed by James Clerk Maxwell over a hundred years ago provided an elegant synthesis of a wide range of phenomena involving electric and magnetic fields. In particular, his theory provided an explanation of the wave propagation of light for it predicted the existence of electromagnetic disturbances that should travel with a speed equal to the accepted value of the speed of light. These disturbances consist of coupled oscillating electric and magnetic fields that are perpendicular to each other and to the direction of propagation of the radiation (Fig. 1.1). The frequency of oscillation and the wavelength of the

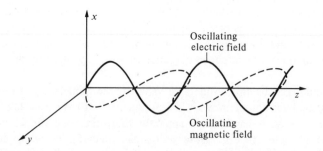

Fig. 1.1. Electromagnetic radiation travelling in the z-direction. The radiation illustrated here is plane polarized as the electric field is always in the xz-plane.

radiation are related by the equation

$$c = \nu\lambda \tag{1.1}$$

where c is the speed of the radiation (commonly known as the speed of light), ν is the frequency, and λ is the wavelength. The positions of spectral lines can be characterized by their frequency, wavelength, or wave number $1/\lambda$, all of which are equivalent as they are related to each other by eqn (1.1). The spectral regions of electromagnetic radiation are shown in Fig. 1.2, from which it can be seen that visible light occupies just a small fraction of the overall range of frequencies.

Electromagnetic radiation can also be regarded as consisting of discrete packets or quanta of energy that travel with the speed of light. The fundamental relationship that links the corpuscular and wave-like nature of radiation is

$$\epsilon = h\nu \tag{1.2}$$

where ϵ is the energy of the quanta, ν is the frequency of the associated radiation, and h is the Planck constant. The interaction of a molecule with radiation involves the absorption or emission by the molecule of a quantum of radiation which is accompanied by a transition from one energy level of the molecule to another. Since energy is conserved, the energy difference ΔE between the two levels must be equal to the energy of the quantum of radiation, i.e.

$$\Delta E = h\nu \tag{1.3}$$

where ν is the frequency of the radiation that is absorbed or emitted. The various forms of spectroscopy can therefore be classified according to the frequency of the radiation or by the type of transition that is involved. For example, the absorption or emission of radiation in the X-ray region (Fig. 1.2) is accompanied by

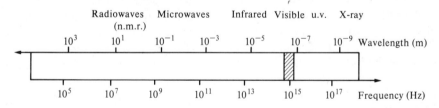

Fig. 1.2. Regions of the electromagnetic spectrum. The visible region occupies a small band around 10^{15} Hz, as shown by the shaded area.

changes in the energy of the inner electrons of atoms or molecules. The visible and ultraviolet regions are associated with transitions of the valence electrons, and in fact can be used to investigate the redox state of living systems. At lower frequencies, the microwave and infrared regions are characteristic of molecular rotational and vibrational energy changes respectively. The frequencies characteristic of n.m.r. lie in the radiofrequency region of the electromagnetic spectrum. They are low, typically in the range 1–500 MHz, and are therefore (from eqn (1.3)) associated with transitions between energy levels that are relatively closely spaced. These levels correspond to different magnetic states of atomic nuclei, as we now consider.

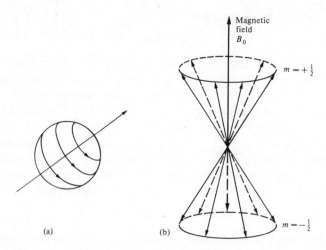

Fig. 1.3. (a) A nucleus can be visualized as spinning about its own axis which is the axis of its magnetic moment (see § 5.1). (b) Orientations that can be taken up in an applied field B_0 by the magnetic moment of a nucleus of spin $\frac{1}{2}$. The orientations are specified by the quantum number m (see § 5.2.1) and describe two cones.

1.3. The basis of n.m.r.

Certain atomic nuclei, such as the hydrogen nucleus ^{1}H or the phosphorus nucleus ^{31}P, possess a property known as spin. This can be visualized as a spinning motion of the nucleus about its own axis (Fig. 1.3(a)). Associated with the spin is a magnetic property, so that the nucleus can be regarded as a tiny bar magnet with its axis along the axis of rotation. If a magnetic field is applied to a sample containing such nuclei (e.g. to water in the case of ^{1}H n.m.r.), we might expect the nuclear magnets to align along the field just as a compass needle aligns along a magnetic field. However, the nuclei have spin and obey the laws of quantum mechanics, and therefore they do not behave like conventional bar magnets. Instead, we find that nuclei such as ^{1}H that have a spin quantum number $I = \frac{1}{2}$ (see § 5.2.1) can have one of two orientations with respect to the applied field, as shown in Fig. 1.3(b)). These two orientations have slightly different energies, and the energy difference between the two states is proportional to the magnitude of the applied field (see Fig. 1.4). Transitions between these states can be induced by applying an oscillating magnetic field[1] of frequency ν_0 that satisfies eqn (1.3), i.e.

$$\Delta E = h\nu_0$$

where ΔE is the energy separation of the levels. We then find (see § 5.2.3) that

$$\nu_0 = \gamma B_0/2\pi \tag{1.4}$$

---

[1] Thus, strictly speaking, n.m.r. does not (as is commonly believed) utilize electromagnetic radiation which, as we have seen, involves coupled electric and magnetic fields; only the magnetic component is present in n.m.r. experiments.

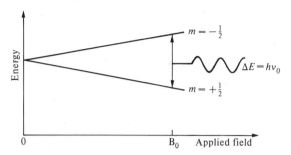

Fig. 1.4. Energies of the two orientations of a nucleus of spin $\frac{1}{2}$ plotted as a function of the applied field B_0.

where B_0 is the magnitude of the applied magnetic field. γ is known as the magnetogyric ratio of the nucleus and ν_0 is the resonance frequency. γ varies from one nuclear isotope to another, and this is why ^{1}H, ^{13}C and ^{31}P n.m.r. for example, are all performed at different frequencies in a given field. (^{1}H n.m.r. is known as proton n.m.r. because the hydrogen nucleus consists of a single proton.)

The nuclear magnets interact very weakly with the applied field B_0, and this accounts for the low value of the energy separation ΔE and of the characteristic n.m.r. frequencies. We shall see in § 5.2.4 that this weakness is also responsible for the low signal intensities obtained in n.m.r. spectra.

1.4. The n.m.r. spectrum

The resonance frequency of a nucleus is directly proportional to the *local* magnetic field experienced by the nucleus. In the preceding section we considered this field to be the applied field B_0. If this were true, all protons for example would absorb energy at the same frequency, and n.m.r. would be a relatively uninteresting and uninformative technique incapable of resolving between protons from different atoms or molecules. However, the applied field B_0 also induces electronic currents in atoms and molecules, and these produce a further small field $B_{0\sigma}$ at the nucleus which is proportional to B_0. The total effective field B_{eff} at the nucleus can therefore be written

$$B_{\text{eff}} = B_0 (1 - \sigma) \tag{1.5}$$

where σ expresses the contribution of the small secondary field generated by the electrons. Using eqn (1.4) we find that

$$\nu_0 = \frac{\gamma}{2\pi} B_0 (1 - \sigma). \tag{1.6}$$

σ is a dimensionless constant, known as the shielding or screening constant, and has values typically in the region $10^{-6} - 10^{-3}$. The magnitude of σ is dependent upon the electronic environment of the nucleus, and therefore nuclei in different chemical environments give rise to signals at different frequencies. The separation of resonance frequencies from an arbitrarily chosen reference frequency is termed the *chemical shift*, and is expressed in terms of the dimensionless units of parts per

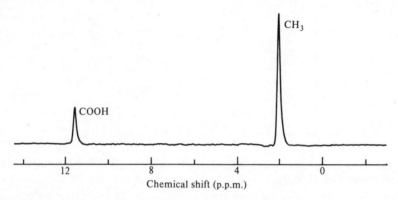

Fig. 1.5. ^{1}H n.m.r spectrum of acetic acid (CH$_3$COOH). The relative areas of the two signals are 3:1. The frequencies of the signals are expressed in terms of p.p.m. relative to the signal from the reference compound tetramethylsilane (TMS).

million (p.p.m.). A simple illustration of the chemical shift is given by the ^{1}H n.m.r. spectrum of acetic acid shown in Fig. 1.5. The CH$_3$ and COOH protons experience different chemical environments and therefore give rise to two separate signals or resonances.

The *intensities* of n.m.r. signals, as measured from their *areas*, are proportional

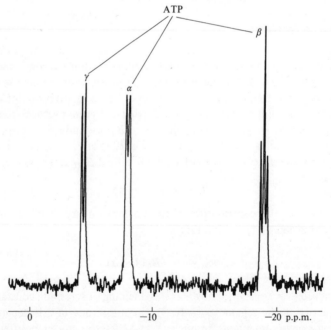

Fig. 1.6. ^{31}P n.m.r. spectrum obtained at 73.8 MHz from a solution of ATP at pH 7.0. The chemical shifts in this spectrum, and in most of the other ^{31}P spectra in this book, are expressed relative to the signal of phosphocreatine at pH 7, which is assigned the value 0 p.p.m. (see § 6.1.2 for a discussion of frequency standards).

to the number of nuclei that contribute towards them, and so in Fig. 1.5 the relative areas of the two signals are 3:1. However, it should be noted that other factors, such as the *spin–lattice relaxation time* T_1 amd the *nuclear Overhauser effect* (to be discussed in Chapter 6), can also affect signal intensities.

The ^{31}P n.m.r. spectrum of adenosine triphosphate (ATP), shown in Fig. 1.6, contains three groups of spectral lines corresponding to the α-, β-, and γ-phosphates of ATP, all of which are chemically different. In addition, each group is split into a multiplet of lines as a result of an interaction between the neighbouring phosphorus nuclei which is known as *spin–spin coupling*. This interaction is transmitted through electronic bonds, and the magnitude of the splitting is independent of the applied field B_0.

Signals from molecules in solution often have a characteristic lineshape $g(\nu)$ given by

$$g(\nu) \propto \frac{T_2}{1 + 4\pi^2\, T_2^2\, (\nu - \nu_0)^2} \tag{1.7}$$

where ν_0 is the resonance frequency, and T_2 is the *spin–spin relaxation time*. This is known as a Lorentzian lineshape, and is illustrated in Fig. 1.7. The intensity of the signal is given by the shaded area. The natural *linewidth* $\Delta\nu_{1/2}$ at half height is given by

$$1/T_2 = \pi\Delta\nu_{1/2}. \tag{1.8}$$

In general, as molecules become increasingly immobilized they produce broader signals, and so, at least in principle, linewidths can provide information about molecular motion. However, linewidths and lineshapes can also be affected by other factors such as *chemical exchange* and magnetic field inhomogeneity.

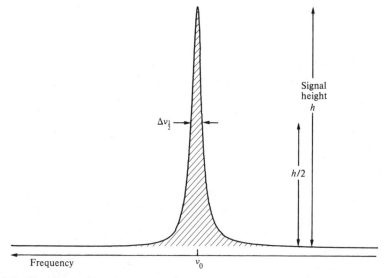

Fig. 1.7. The Lorentzian lineshape. $\Delta\nu_{1/2}$ is the natural linewidth determined by T_2; the observed linewidth is often greater because of effects such as B_0 inhomogeneity. The intensity of the signal is given by the shaded area, and the signal height, or amplitude, is equal to h.

All of the parameters discussed above can yield information about some aspect of the structure, chemical environment, mobility, concentration and reaction rates of the molecules under investigation. They are discussed in detail in Chapter 6, but an idea of the way in which n.m.r. spectra can be interpreted can be obtained from examples given later in this introductory chapter.

1.5. The n.m.r. nuclei

Only those nuclei that have magnetic properties, i.e. those that possess non-zero spin (see Chapter 5), give rise to n.m.r. signals. Some of these nuclei, together with their n.m.r. properties, are listed in Table 1.1. It should be noted that nuclei of spin

Table 1.1
N.m.r. properties of the nuclei commonly used in biology.

Nucleus	Spin quantum number	Resonance frequency at 5T, in MHz	Natural abundance (%)	Relative sensitivity at constant field[1]
^{1}H	1/2	213.0	99.98	100
^{2}D	1	32.7	0.0156	1.5×10^{-4}
^{13}C	1/2	53.5	1.1	1.6×10^{-2}
^{14}N	1	15.4	99.6	1.0×10^{-1}
^{15}N	1/2	21.6	0.36	3.7×10^{-4}
^{19}F	1/2	200.0	100.0	83.0
^{23}Na	3/2	56.3	100.0	9.3
^{31}P	1/2	86.2	100.0	6.6
^{35}Cl	3/2	20.9	75.4	3.5×10^{-1}
^{39}K	3/2	9.9	9.1	4.6×10^{-2}

[1] Relative sensitivity is the n.m.r. sensitivity of the nucleus relative to that of an equal number of protons, multiplied by the percentage natural abundance.

greater than $\frac{1}{2}$ tend to give broader signals than those of spin $\frac{1}{2}$, and therefore high-resolution n.m.r. experiments (i.e. those experiments in which signals of different frequencies can be resolved from each other) generally utilize nuclei of spin $\frac{1}{2}$ such as ^{1}H, ^{13}C, and ^{31}P. These have proved so far to be the most informative nuclei in biological n.m.r. The abundant isotopes of carbon and oxygen, ^{12}C and ^{16}O, have zero spin and therefore do not produce n.m.r. signals.

1.6 Sample requirements: sensitivity and resolution

It is essential that samples produce signals that are intense enough to be distinguishable from noise and narrow enough to be distinguishable from each other. These are the familiar problems to the n.m.r. spectroscopist of sensitivity and resolution.

Narrow signals are usually obtained only from molecules that are fairly mobile, and therefore most high-resolution n.m.r. studies are performed on solutions. Many of the metabolites in biological tissue are freely mobile, and therefore it is not surprising to find that intact living systems can give rise to high resolution spectra. For example, the ^{31}P n.m.r. spectra of frog gastrocnemius muscles shown in Fig. 1.8 contain resonances from ATP, phosphocreatine, and inorganic phosphate. However,

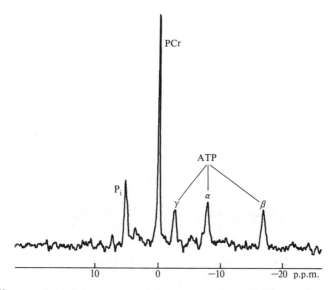

Fig. 1.8. ^{31}P n.m.r. spectrum of frog gastrocnemius muscles. The spectrum was obtained at 73.8 MHz in about 1 min from intact muscles weighing about 2 g *in toto*. PCr refers to phosphocreatine, and P_i to inorganic phosphate, which is at a fairly high concentration because the muscles had been maintained under anaerobic conditions for several minutes at 20 °C. The spectrum was obtained by Dawson, Gadian, and Wilkie using the probe described in § 8.8.2.

one would not expect to detect high-resolution signals from the phosphorus of deoxyribonucleic acid (DNA) or phospholipids, as these compounds, being highly immobilized, would produce very broad signals, the width of which might be similar to or greater than the full spectral width shown in Fig. 1.8.

A major disadvantage of n.m.r. is its inherent lack of sensitivity which can be expressed in terms of the signal-to-noise ratio of the spectral lines:

$$\text{signal-to-noise ratio} = \frac{2.5 \times \text{signal height}}{\text{peak-to-peak noise}}$$

(see Fig. 1.9). The signal-to-noise ratio is dependent upon a wide range of factors including (i) the nucleus that is being studied (see Table 1.1 for relative sensitivities), (ii) the volume of sample, (iii) the magnetic field strength B_0 of the spectrometer, (iv) the design of the spectrometer, particularly of the radiofrequency coil (see Chapter 8), (v) the time for which the spectra are accumulated, (vi) the splitting and width of the spectral lines, (vii) the relaxation times T_1 and T_2, (viii) the extent of isotopic enrichment where this is feasible, and of course (ix) the concentration of the nuclei under investigation.

Because of the large number of variables, it is impossible to give anything other than an order-of-magnitude estimate for the concentration of sample that is required. However, if one wishes to obtain reasonable spectra in a few minutes, one can expect to require about 0.4 ml of a 0.3 mM sample for ^{1}H n.m.r., 1 ml of a

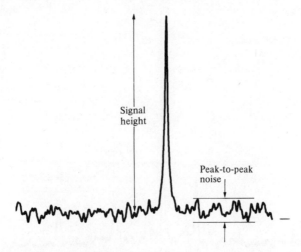

Fig. 1.9. Signal and noise in an n.m.r. spectrum.

1 mM sample for ^{31}P n.m.r., and 2 ml of a 1 mM sample for ^{13}C n.m.r. In the case of ^{13}C n.m.r. 1 mM refers to the concentration of the ^{13}C isotope; if isotopic enrichment is not possible a concentration of 100 mM would be required.

The volume of sample is often fixed by the design of the spectrometer. For example, ^{1}H n.m.r. experiments are often performed on about 0.5 ml of sample in a tube of outer diameter 5 mm. The maximum volume that one can expect to use in conventional ^{13}C or ^{31}P n.m.r. experiments is about 10–20 ml. However, the advent of magnets of very wide bore permits the use of much larger volumes, so that it is now feasible to obtain informative spectra not only from intact animals, such as rats and rabbits, but also from human beings.

1.7. Detection of n.m.r. signals

1.7.1. The spectrometer

Figure 1.10 shows a very simple block diagram of the instrumentation required for an n.m.r. spectrometer. The transmitter transmits radiation of the appropriate fre-

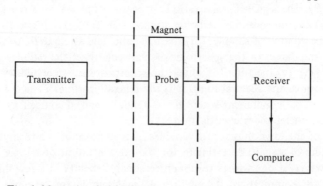

Fig. 1.10. A simple block diagram of an n.m.r. spectrometer.

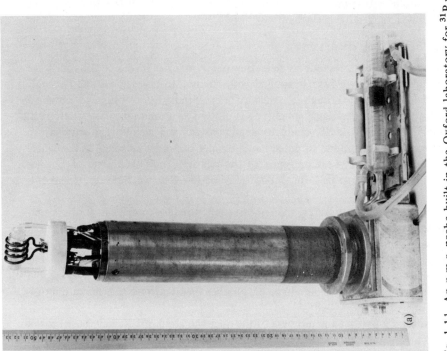

Fig. 1.11. An n.m.r. probe built in the Oxford laboratory for ^{31}P n.m.r. studies of perfused rat hearts (see Garlick, Radda, and Seeley 1979). (a) The complete probe, except that the top half of the surrounding aluminium alloy tube has been removed to reveal the sample chamber and radiofrequency coil; (b) a close-up showing in more detail 1, the sample chamber; 2, the coil; 3, the tuning capacitor; 4, the perfusion fluid inflow and outflow, and 5, the thermostatting air jets. The rulers are present as an indication of scale (in cm). It should be pointed out that this is not a conventional type of probe; it was designed specifically for studying rat hearts (see § 8.8.3).

quency into what is called the probe. The probe (see Fig. 1.11) contains the sample and is situated in the strong uniform magnetic field B_0. The sample absorbs radio-frequency energy, and the ensuing signal is fed into the receiver and then into the computer where it is processed in the required way.

The main feature of the probe is a coil of wire that surrounds the sample. This coil transmits power into the sample in the form of a radiofrequency magnetic field B_1, the frequency of which corresponds to the resonance frequency of the nuclei under investigation. The radiofrequency coil also detects the signal associated with the absorption of energy by the nuclei within the sample.

A typical n.m.r. spectrometer for use in high-resolution studies employs a super-conducting magnet that produces a field of about 6 T. It is essential that the field be extremely uniform, or homogeneous, over the sample volume, for it can be seen from eqn (1.4) that any inhomogeneity ΔB_0 will broaden resonances by an amount $\Delta \nu = (\gamma/2\pi)\Delta B_0$. In addition to impairing resolution, this broadening can also reduce the information about molecular structure and mobility that can be obtained from resonance linewidths. It is therefore desirable that field inhomogeneity should not produce the dominant contribution to resonance linewidths. Since many proton resonances at 270 MHz have natural linewidths in the region of 1 Hz, field homogeneity as remarkable as 1 part in 10^9 is often required. This can be obtained if the sample is spun sufficiently rapidly (about 20 Hz) to average out most of the residual inhomogeneity. The requirements for most studies of living systems are not so stringent; this is indeed fortunate as often it is neither desirable nor feasible to spin a living system.

1.7.2. The modes of detection

There are two main methods of detecting n.m.r. signals. In the continuous-wave mode radiofrequency power is applied continuously to the sample and the magnetic field is swept through a range of field strengths in order to obtain a spectrum in which the signal amplitude is plotted as a function of field strength. (Alternatively, the field B_0 could be kept constant and a radiofrequency sweep used.) However, this mode of detection is rapidly becoming obsolete in biological n.m.r. as a result of the emergence of the relatively new technique of Fourier transform n.m.r. It is this latter mode of detection that we shall discuss in this book.

In Fourier transform n.m.r. the radiofrequency field is applied in short power-ful pulses, typically of duration about 20 μs, the bandwidth of which (i.e. spread in frequency) is sufficiently large to excite *all* of the resonances. Its main advantage over the continuous-wave method is that it provides a considerable improvement in sensitivity (i.e. in the signal-to-noise ratio of the spectral lines) because the resonances are all detected simultaneously rather than one by one as in continuous-wave n.m.r.

It can be shown that the response of the nuclei to a radiofrequency pulse bears a fixed mathematical relationship to the more conventional absorption spectrum obtained in continuous-wave n.m.r. Therefore, by applying to the response the required mathematical manipulation, which is known as Fourier transformation, a

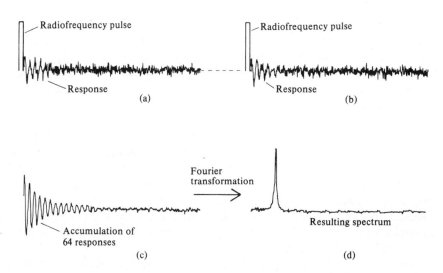

Fig. 1.12. Detection of signals using Fourier transform n.m.r. This diagram illustrates the accumulation of a spectrum containing a single resonance. (a) and (b) are two consecutive responses to radiofrequency pulses. Typically, the pulses are of about 20 μs duration and are applied at intervals of about 1 s. (c) is the accumulation of 64 responses or scans (scaled down from (a) and (b) by a factor of 16) and (d) is the spectrum obtained on Fourier transformation of the accumulation. (From Gadian 1977.)

conventional n.m.r. spectrum can be obtained. Unfortunately, because biological molecules are present at low concentrations they produce weak signals, and so it is almost always necessary to improve the signal-to-noise ratio of the spectral lines by adding together a large number of responses. The accumulation of N responses (which are often rather loosely called scans) leads to an improvement of $\sqrt{N}$, because the signal increases by a factor of N whereas the noise, being random, increases by $\sqrt{N}$. Thus in Fourier transform n.m.r. (see Fig. 1.12) a radiofrequency pulse is applied, the response is observed, and the process is repeated a number of times at intervals of typically 1 s. (The pulse interval is dependent upon the values of the spin–lattice relaxation times (see § 7.2.2)). The responses are automatically added in a computer until the required signal-to-noise ratio is obtained, and then Fourier transformation of the data is performed using this computer to produce the final spectrum.

The dependence of the signal-to-noise ratio on $\sqrt{N}$ means that an improvement of, say, tenfold in the signal-to-noise ratio requires an increase in data accumulation time of a factor of 100. Expressed slightly differently, if a 5 mM solution of ATP produces the required signal-to-noise ration in 1 min, a 500 μM solution would require 100 min of accumulation time.

1.8. Applications of n.m.r. in biology

N.m.r. can be used in a wide variety of ways, but it will be convenient for us to distinguish between three different types of application.

1.8.1. Structure and function of macromolecules

Since the early 1950s n.m.r. has been recognized as a powerful means of studying the structure of molecules in solution, and the technique has proved to be of great value both to the organic chemist and more recently to the biochemist. The scope of the method can be illustrated by experiments that have been performed on the enzyme lysozyme (Dobson 1977; Campbell and Dobson 1979).

An important conclusion of the work on lysozyme is that the solution structure of the enzyme as determined by n.m.r. is similar to the crystal structure determined by X-ray crystallography. In addition, the n.m.r. studies demonstrate that some groups in the protein have extensive mobility, and this correlates well with lack of definition in certain regions of the electron-density map. N.m.r. is far better suited than X-ray crystallography for monitoring conformational changes resulting from chemical modification, binding of ligands, etc., and it does appear that certain conformational changes of lysozyme that are observed in solution may differ from those detected in the crystal.

N.m.r. studies of macromolecules have been extensively reviewed over the last few years (see, for example, Campbell and Dobson 1979). Therefore, although in this book we shall discuss some of the n.m.r. techniques on which these studies rely, we shall not review the important contribution that n.m.r. has made to our understanding of molecular structure and function.

1.8.2. Metabolism

An alternative more empirical approach is to use n.m.r. as a means of identifying the presence of particular molecules within a sample. This can be regarded as the 'molecular fingerprint' approach for it relies on the empirical observation that different molecules produce their own characteristic spectra. Once the molecules have been identified, changes in the intensities of their spectral lines can be interpreted in terms of the kinetics of reactions in which they are involved. In addition, the precise nature of the spectra can yield further information about the detailed structure of the molecules or about their chemical environment. Most studies of metabolism within living systems adopt this approach, which is perhaps fortunate, because interpretation of the spectra is generally more straightforward than it is for structural studies.

1.8.3. Imaging

A third approach involves the use of magnetic field gradients to provide anatomical images of intact biological systems. This relatively new n.m.r. method has been termed 'spin-imaging' or 'zeugmatography', and is discussed in Chapter 4 where we also discuss how n.m.r. can be used to study the metabolic state of selected, localized regions within whole animals and human beings.

1.9. N.m.r. studies of living systems

1.9.1. A historical perspective

The idea of using n.m.r. to study living systems is by no means new. In fact, very soon after the first successful n.mr. experiments (Bloch, Hansen, and Packard 1946;

Purcell, Torrey, and Pound 1946), Bloch obtained a strong proton signal on placing his finger in the radiofrequency coil of his spectrometer. Four years later Shaw and Elsken (1950) used ^{1}H n.m.r. to investigate the water content of potato and maple wood. On the basis of their observations they suggested that n.m.r. might provide a useful method for the rapid determination of the water content of hygroscopic materials. Odeblad and Lindstrom (1955) obtained low-resolution ^{1}H n.m.r. signals from a number of mammalian preparations, including human red blood cells and striated muscle and fat tissue of the rat. Singer (1959) used ^{1}H n.m.r. to measure blood flow in the tails of mice, and he suggested the possibility of making similar measurements of blood flow in human beings.

In the 1962 edition of his book *Physical Chemistry*, Moore posed the prophetic question:

> Suppose a biochemist friend who professed to be quite ignorant about NMR comes to your for advice. He would like to study the transformation ATP → ADP which occurs in muscle metabolism by monitoring the NMR of the ^{31}P nuclei. He is rather wealthy, and has a complete NMR apparatus, including oscillators at 40, 30, 20, 10 and 3 mc/sec, and a magnet which will go up to about 12 000 gauss. Which oscillator would you suggest he use to study the ^{31}P spectra and why? Calculate the value of the magnetic field at which the resonances would appear.

Clearly, the potential of n.m.r. in biology has been appreciated for many years, but unfortunately the early experiments were limited in scope by the relatively poor quality of the instrumentation that was available at the time. For example, the first 'high-resolution' ^{1}H n.m.r. spectrum of a protein had been reported by Saunders, Wishnia, and Kirkwood (1957) who were only able to resolve four broad peaks from the enzyme ribonuclease A.

The development of high-field superconducting magnets in the late 1960s together with the emergence of Fourier transform n.m.r. revolutionized the scope of n.m.r. and made a rapid impact on the study of proteins and other biological molecules. Rather surprisingly, it emerged much more slowly that n.m.r. might have extensive applications in the study of the metabolism of living systems.

Then Moon and Richards (1973) reported high-resolution ^{31}P n.m.r. studies of intact red blood cells. They detected signals from 2,3-diphosphoglycerate, inorganic phosphate, and ATP, and showed how the spectra could be used to determine intracellular pH. At about this time it was also shown that ^{13}C n.m.r. could be used to follow the end products of metabolic pathways (Eakin, Morgan, Gregg, and Matwiyoff 1972; Séquin and Scott 1974). Meanwhile Damadian (1971) had shown that certain malignant tumours of rats differed from normal tissues in their ^{1}H n.m.r. properties, and suggested that ^{1}H n.m.r. might therefore have diagnostic value. Then Lauterbur (1973) coined the word 'zeugmatography' to describe an n.m.r. method for studying the spatial distribution of molecules within a sample.

In 1974 it was reported that high-resolution ^{31}P n.m.r. spectra could be obtained from intact muscle freshly excised from the hind leg of the rat (Hoult *et al.* 1974). The spectra were similar in form to that shown in Fig. 1.8, and the signals could be assigned to the three phosphate groups of ATP, to phosphocreatine, and to inorganic phosphate. As demonstrated by Moon and Richards (1973) in their studies of red cells the chemical shift of the inorganic phosphate resonance defines the intracellular pH, and for the freshly excised muscle the pH was measured to be 7.1. In addition, the chemical shifts of the three ATP signals are sensitive to the binding of Mg^{2+} ions, as first shown by Cohn and Hughes (1962). The values observed in the muscle spectra indicated that the ATP in muscle is predominantly complexed to Mg^{2+} ions. In these early experiments the muscles were not maintained under controlled physiological conditions. Therefore, as expected, there was a gradual breakdown of phosphocreatine followed by depletion of ATP together with a fall in intracellular pH resulting from the accumulation of lactic acid. Similar observations were reported by Burt, Glonek, and Bárány (1976a) for other muscle types.

The studies discussed above established the basis for subsequent work in which great care has had to be taken to ensure that all preparations are maintained under controlled physiological conditions within the n.m.r. spectrometer. This recent work is discussed in detail in later chapters, but we now outline the rapid progress that has been made by discussing a few examples from the recent literature in the field.

1.9.2. Some recent studies

Following on from the experiments of Hoult *et al.* (1974), it was shown that high-resolution ^{31}P n.m.r. spectra could be obtained from frog sartorius muscles that were maintained in good physiological condition within the spectrometer (see § 3.6). It proved possible to stimulate the muscles electrically, measure their tension response, and follow the resulting changes in their ^{31}P n.m.r spectra. Experiments on other tissues and organs soon followed, and a wide range of studies on perfused isolated hearts, kidneys, and livers have now been performed, some of which are discussed in Chapter 3. More recently, developments in magnet technology and in radiofrequency coil design have permitted high-resolution ^{31}P n.m.r. spectra to be obtained from selected regions of live animals such as the rat. In these experiments (Ackerman, Grove, Wong, Gadian, and Radda 1980b) an anaesthetized rat is positioned within the magnet and a radiofrequency coil is placed in the required position against a leg of the rat. Remarkably clear ^{31}P n.m.r. signals can be detected (see Fig. 1.13(a)) from a disc-shaped region of the leg immediately in front of the coil; in the studies described here this 'sensitive' region has a volume of about 1 ml. It should be stressed that the animal undergoes no surgery whatsoever; the experiment can be likened to examining the condition of the heart with a stethoscope.

The resonances in Fig. 1.13(a) arise from ATP, phosphocreatine, and inorganic phosphate within the leg muscle; the lack of signal from 2,3-diphosphoglycerate indicates that any blood within the tissue makes a negligible contribution to the spectrum, and the contribution from skin is also very small. The chemical shifts of the ATP resonances confirm, as expected, that the ATP in the muscle is predominantly complexed to divalent metal ions which are assumed to be Mg^{2+}. In addition, from the chemical shift of the inorganic phosphate signal the intracellular pH is estimated to be 7.1.

From the areas of the various signals interesting conclusions can be reached

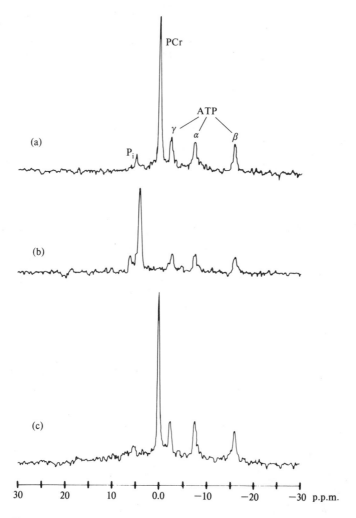

Fig. 1.13. ^{31}P n.m.r. spectra obtained at 73.8 MHz from the leg of an anaesthetized rat. Each spectrum represents the accumulation of 128 scans repeated at 12 s intervals. The probe used for this work was similar to that shown in Fig. 8.10. For further details, see text. (From Ackerman *et al.* 1980*b*. Reprinted by permission from *Nature* **283**, 167–70. Copyright © 1980 Macmillan Journals Limited.)

about the relative concentrations of the metabolites. In particular, the concentration of inorganic phosphate derived from these studies is very much lower than that measured by freeze-clamping techniques. Comparisons such as this have a number of interesting kinetic and thermodynamic implications and will be discussed in detail in § 3.1.

The spatial selectivity of the method makes it possible to assess the effects of localized ischaemia (cessation of blood flow) within the rat leg muscle. Figure 1.13(a) shows a spectrum obtained from muscle below the knee under normal conditions. Figure 1.13(b) shows a spectrum obtained from the same location after a tight tourniquet had been placed around the leg just above the knee. There are dramatic differences between the two spectra, for the spectrum from the ischaemic muscle shows that there is very little phosphocreatine, somewhat reduced amounts of ATP, and a very large concentration of inorganic phosphate. Furthermore, the chemical shift of the inorganic phosphate resonance indicates that the pH had declined to 6.7. A spectrum obtained from leg muscle above the tourniquet (Fig. 1.13(c)) is very similar to that shown in Fig. 1.13(a), confirming that the muscle above the knee has not suffered from ischaemia. Similar studies of human limbs are now feasible (see Chapter 4), and it is clear that n.m.r. could prove to be of considerable value in investigating the biochemistry of healthy and diseased human muscle.

[31]P n.m.r. therefore provides an elegant non-invasive means of assessing the energetic state of living systems, and a variety of studies are described in Chapters 3 and 4. However, it is not particularly well suited to studying the intricate details of metabolic pathways, because relatively few phosphorylated intermediates are present in sufficiently high concentrations to generate measurable signals. An alternative approach which is looking increasingly attractive is to use [13]C n.m.r. to follow the metabolism of a ^{13}C label that is introduced into the living system. This approach is well illustrated by some interesting studies of metabolism in *Escherichia coli* (Ugurbil *et al.* 1978*a*).

Figure 1.14 shows a [13]C n.m.r. spectrum obtained 5 min after adding glucose enriched with ^{13}C at the C-1 position to an anaerobic suspension of *E. coli*. The signals can be assigned to (from left to right) the labelled carbon of β- and α-glucose, carbon at the C-1 and C-6 positions of fructose 1,6-diphosphate, buffer, malate, succinate, lactate, ethanol, and alanine (see figure caption for more details).

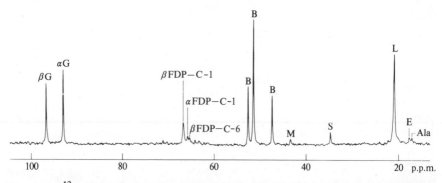

Fig. 1.14. [13]C n.m.r. spectrum obtained in 1 min at 90.5 MHz from a suspension of *E. coli* at 20 °C. βG and αG are the signals from the ^{13}C at the C-1 position of the two anomers of glucose; the peaks labelled B are from the buffer, M is from malate, S is from the C-2 position of succinate, L is from the C-3 position of lactate, E is from the methyl carbon of ethanol, and Ala is from the C-3 position of alanine. The peaks labelled FDP are from the C-1 and C-6 positions of fructose 1,6-diphosphate. (From Shulman *et al.* 1979. Copyright © 1979 by the American Association for the Advancement of Science.)

The assignments are made on the basis of chemical shifts observed both in cells and extracts, together with the use of enzymatic digestions and common sense. The signal intensities observed just after the addition of glucose were consistent with the β- and α-anomers of glucose being in equilibrium with a $\beta:\alpha$ ratio of 64:36. However, the spectrum of Fig. 1.14 shows that the β-anomer has a lower concentration than the α-anomer, indicating that this particular strain of *E. coli*, MRE 600, preferentially catabolizes the β-anomer and does not have high anomerase activity. The main product of glycolysis is seen to be lactate, but signals from succinate, malate, and ethanol also appear to lesser extents.

The signals in the fructose 1,6-diphosphate region of the spectrum are particularly interesting. They indicate that some of the ^{13}C label, which would be expected to flow into the C-1 position of fructose 1,6-diphosphate, ends up in the C-6 position. Exchange of these two carbons can occur via the reaction catalysed by triose phosphate isomerase (see Fig. 3.1), and the relative intensities of the signals from the two carbons can therefore provide information about the metabolic flux through some of the glycolytic enzymes. For example, if the reactions catalysed by triose phosphate isomerase and aldolase were close to equilibrium, one would expect the 1- and 6-carbons of fructose 1,6-diphosphate to be equally labelled. From these studies, together with further experiments using glucose labelled at the C-6 position, it was possible to conclude that there is considerably more metabolic flux through the aldolase reaction towards the triose phosphates than there is in the reverse direction.

It should be stressed that the main strength of metabolic n.m.r. studies, as illustrated by the examples given above and in later chapters, does not lie in elucidating *which* reactions take place in living systems, for there is already a vast array of knowledge on the subject that has been built up over the years. What is more important is that n.m.r. provides a method of monitoring reaction *rates* and enzyme activities *in vivo*. By relating these observations to metabolite levels and physiological function, it should be possible to develop a more comprehensive picture of how metabolic pathways and physiological activity are controlled and integrated.

The final example in this section describes the use of ^{1}H n.m.r. to obtain images of biological material. Plate 1(a) shows the distribution of mobile protons in a thin section through a human head, obtained by n.m.r. spin-imaging. The principles of the method are described in Chapter 4, but here we simply point out that the signals that constitute the image arise from mobile protons contained in water and lipids; mobility here refers to molecular tumbling motion, rather than to processes such as flow that represent macroscopic movement through space.

The head scan shown in Plate 1(a) correlates well with the expected structure shown in Plate 1(b), and there are several interesting features of the image. For example, the skull contains very little mobile water, and its consequent lack of signal (dark areas indicate low signal) enables the surface of the brain to be seen clearly. This contrasts with X-ray computed tomography (CT) scans, for which absorption by bone can generate artefacts and cause blurring of features close to the skull. The resolution of the method is illustrated by the contrast between the retro-orbital fat pads and the optic nerves.

^{1}H spin-imaging is still in the early stages of development, but it is now apparent that the method could complement existing medical imaging methods such as ultrasonic imaging and X-ray CT scanning. In comparing n.m.r. with other imaging methods, it should be noted that n.m.r. is non-invasive, and does not involve ionizing radiation; in fact it is believed that most, if not all, methods of n.m.r. imaging are free from hazard. Spin-imaging and its applications are discussed in much greater detail in Chapter 4.

1.10. The relative merits of ^{1}H, ^{13}C, and ^{31}P n.m.r.

The examples described in the previous section illustrate some of the possibilities afforded by ^{1}H, ^{13}C, and ^{31}P n.m.r., and in this final section of the chapter, we discuss the relative merits of these three nuclei.

The proton is the most sensitive nucleus, in that it produces a greater signal-to-noise ratio than any other nucleus (apart from tritium) in a given period of time. The two distinctive features of ^{1}H n.m.r. are the complexity of the spectra, and the presence of a large solvent peak in spectra of aqueous solutions. These two features have important advantages and disadvantages, as we shall now observe.

Firstly, the existence of a large concentration of water (55 M) makes it possible to perform imaging studies of the type described above and in Chapter 4. Images of this quality and spatial resolution could not possibly be obtained from compounds present at the typical concentrations ($\lesssim 5$ mM) found in living systems. However, the presence of a large solvent peak can create difficulties if we wish to observe the weaker signals from other compounds within the sample. For this reason, high resolution ^{1}H n.m.r. studies are frequently performed using deuterium oxide (D_2O) rather than H_2O as the solvent. Even if D_2O is used, there may still be a large signal from the residual HDO within the sample. Fortunately, various

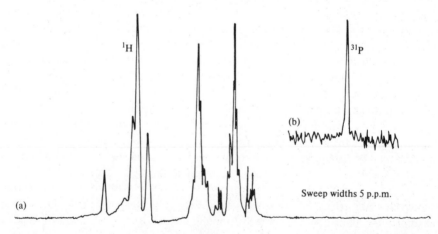

Fig. 1.15. (a) ^{1}H n.m.r. spectrum of glucose 6-phosphate at 270 MHz, and (b) ^{31}P spectrum of the same molecule at 73.8 MHz. The total width of each spectrum is 5 p.p.m. and the linewidths are comparable to values that this molecule might produce when present in intact biological systems. (From Gadian, Radda, Richards, and Seeley 1979.)

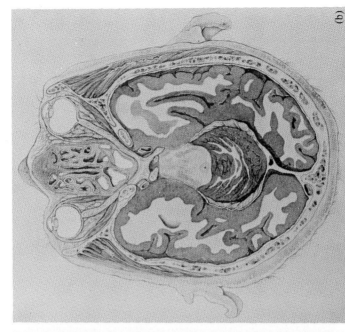

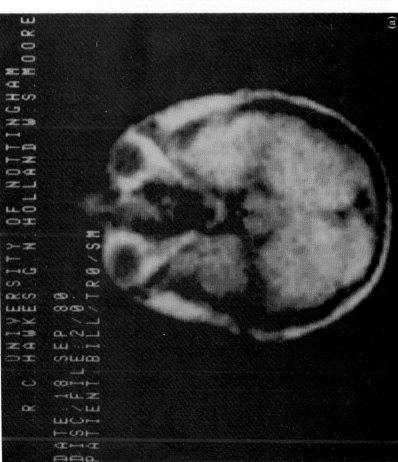

Plate 1. (a) A head scan obtained by ^{1}H n.m.r. spin imaging (from the work of Hawkes, Holland, and Moore, Nottingham University); for details see text and § 4.2.5. (b) For comparison with (a), a diagram illustrating the expected structural features. (From Carter *et al.* 1977.)

methods are now available for suppressing the solvent peak (see § 7.2.6), and under certain circumstances these methods are so effective that it becomes unnecessary to replace H_2O by D_2O.

The second feature of ^1H n.m.r. is that the spectra are very complex, owing to the ubiquity of hydrogen atoms in biological molecules. This is illustrated by Fig. 1.15(a), which shows the ^1H n.m.r. spectrum of glucose 6-phosphate, which is a relatively small molecule. These complex spectra can, of course, be utilized to provide information about the structure and mobility of biological molecules, and ^1H n.m.r. has produced a wide range of important conclusions about the conformation of biological molecules in solution. However, the ^1H spectra of intact biological systems contain overlapping signals from so many compounds that the question must be raised as to whether any detailed interpretation of the spectra is possible. Fortunately, the spectra can often be simplified by means of n.m.r. techniques of varying degrees of sophistication, as illustrated by the studies of red cells described later in the book, and high resolution ^1H n.m.r. studies of cellular suspensions have a wide range of interesting possibilities. However, the problems associated with spectral complexity and the large solvent peak may be more difficult to overcome for studies of relatively large tissue preparations, and it remains to be seen whether high resolution ^1H n.m.r. will prove to be as valuable as ^{13}C or ^{31}P n.m.r. as a method of studying metabolism in whole tissues and organs.

The ^{31}P nucleus is particularly suitable for studies of living systems for the following reasons.

(i) Narrow resonances can be obtained, and they occupy a fairly wide range of chemical shifts (about 30 p.p.m. for biological phosphates).

(ii) Although the sensitivity of ^{31}P n.m.r. is only one-fifteenth of that of ^1H n.m.r., it is nevertheless one of the most sensitive nuclei. Moreover, ^{31}P is the naturally occurring isotope of phosphorus and so no isotopic enrichment is necessary.

(iii) The spectra are simple and far more easily interpreted than ^1H n.m.r. spectra. This is illustrated by the ^{31}P n.m.r. spectrum of glucose 6-phosphate which is shown alongside the ^1H spectrum of the same molecule in Fig. 1.15(b). (However, the simplicity of the ^{31}P spectrum means that it does not provide as clear a 'fingerprint' of the molecule as the ^1H spectrum; other compounds may produce very similar ^{31}P spectra.)

(iv) Several important phosphorus-containing compounds, such as ATP, phosphocreatine, and inorganic phosphate, occur in living systems at concentrations high enough to be detectable by ^{31}P n.m.r. (about 0.2 mM and above). Because of the involvement of these compounds in the energetics of living systems, ^{31}P n.m.r. should provide an ideal means of monitoring the energetic state of living cells, tissues, and organs.

^{13}C has a natural abundance of only 1.1 per cent and therefore it takes a very long time to accumulate spectra in the absence of isotopic enrichment. As a result enrichment is almost invariably essential for studies of living systems, but as long as this is feasible ^{13}C n.m.r. has many attractions. The signals are narrow and occupy a wide range of chemical shifts (about 300 p.p.m.), and therefore spectral resolution is often much better than for ^{1}H n.m.r. In contrast to ^{31}P n.m.r., a wide range of compounds can be detected. Moreover, the metabolic pathway to be investigated can be selected by choosing the appropriate compound for enrichment. Unfortunately, the preparation of ^{13}C-labelled compounds can be difficult, and it may be expensive to buy enriched compounds in the quantities required for n.m.r.

In this book, we consider almost exclusively the ^{1}H, ^{13}C, and ^{31}P nuclei, as it seems likely that these will remain the most popular and informative nuclei in biological n.m.r. Nuclei of spin greater than $\frac{1}{2}$ tend to give broader signals and rarely generate high-resolution spectra. As a result the information that they can provide is usually rather limited, but in § 4.2.2 we do mention one interesting study using the ^{23}Na nucleus which has a spin of 3/2. Another nucleus of spin $\frac{1}{2}$ that has become increasingly popular in recent years is ^{15}N. In comparing carbon and nitrogen metabolism, it is interesting to note that, whereas a wide range of metabolic studies can be performed using ^{14}C as a radioactive tracer, there is no isotope of nitrogen that is suitable for radioactivity experiments. Therefore, metabolic studies using ^{15}N n.m.r. could in principle be of great value. The main problem to overcome is one of sensitivity; the inherent sensitivity of ^{15}N n.m.r. is much worse than that for ^{13}C, and furthermore ^{15}N has a natural abundance of only 0.36 per cent.

Finally, it should be noted that studies involving the ^{1}H, ^{13}C, and ^{31}P nuclei are not mutually exclusive; it is now possible to detect two or more nuclei simultaneously (Styles, Grathwohl, and Brown 1979). The idea of simultaneous detection is particularly attractive because of the complementary nature of the information that is available from the different nuclei, and one can anticipate that this will become a standard feature of future n.m.r. spectrometers.

2

The type of information available from n.m.r.

High-resolution n.m.r. spectra are extremely rich in information, and much of the skill associated with n.m.r. experiments is involved with interpretation of the spectra. The purpose of this chapter is to describe the type of biochemical information provided by n.m.r. studies of living systems and the manner in which this information is extracted from the spectra.

For metabolic studies of living systems n.m.r. has two important advantages over other biochemical techniques. Firstly, it is non-invasive, and the time course of metabolic reactions can therefore be followed in a single experiment performed on a single preparation. This contrasts sharply with techniques that involve freezing of the tissue to stop chemical reactions at a specified time. In these freeze-clamping experiments different specimens are required for each point in time and sufficient data must be obtained to permit statistical analysis to be performed. This can involve a great deal of work and the sacrifice of a large number of experimental animals, and moreover serious problems can arise as a result of variations between batches of animals.

Because n.m.r. is non-invasive, it can also provide information about the intracellular environment of metabolites which is lost if the tissue must be destroyed for chemical analysis. For example, ^{31}P n.m.r. gives information about intracellular pH and the binding of Mg^{2+} ions to ATP.

The second important advantage of n.m.r. is that it is non-specific, in the sense that resonances are observed from all mobile compounds that are present at sufficiently high concentrations. The simultaneous observation of all of these compounds, rather than just those selected for chemical analysis, could permit the detection of metabolism not predicted by current theories and which might be missed by more conventional methods.

The main disadvantage of n.m.r. is that only those compounds that are present at high concentrations (above about 0.2 mM) generate detectable signals. However, as we shall see in this chapter, the information available from n.m.r. studies is not restricted to these compounds alone; important deductions can be made about the concentrations of additional species that cannot be detected directly.

2.1 Identification of resonances

The first stage in interpretation of any spectrum must be to identify the molecular groupings that give rise to the observed resonances. Unambiguous identification can sometimes be trivial, but alternatively can be extremely difficult, if not impossible. For example, ^1H spectra of proteins may contain 1000 or more resonances, and assignment of these to specific amino acid residues is an arduous task (see Campbell and Dobson 1979). In contrast, the ^{31}P n.m.r. spectra of living tissues and organs are usually very simple, and assignment of most of the signals is straightforward. It is ironic that the assignment problem should be so much more severe for a purified protein solution than it is for a whole tissue. The reason for this is, of course, that the comparison is between ^1H spectra in the case of the protein studies and ^{31}P

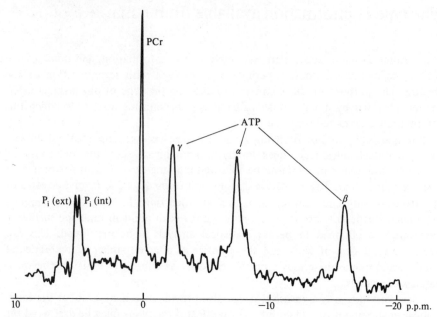

Fig. 2.1. [31]P n.m.r. spectrum obtained in 8 minutes at 73.8 MHz from a Langen-dorff-perfused rat heart. The signals are assigned as shown, and as explained in the text. PCr refers to phosphocreatine, and P_i(int) and P_i(ext) to inorganic phosphate in the intracellular space and the perfusion medium respectively. The chemical shifts in this spectrum are expressed relative to the phosphocreatine signal. (From Grove, Ackerman, Radda, and Bore, 1980.)

spectra in the case of the whole tissue. [1]H spectra are generally far more com-plicated because any macromolecule contains enormous numbers of protons in a range of different chemical environments.

The various methods of assignment are best illustrated by a few examples. Con-sider firstly the [31]P n.m.r. spectrum of a perfused rat heart shown in Fig. 2.1. The signal at about −16 p.p.m. is well separated from the other signals and has a chemical shift characteristic of the β-phosphate group of nucleoside triphosphates. On the basis of metabolite concentrations known to exist in the heart it is incon-ceivable that a significant contribution to this signal could be generated by any compound other than ATP, and therefore this signal can be assigned to the β-phos-phate of ATP. Having made this assignment, the precise value of the chemical shift can then be used to provide information about the intracellular environment of ATP as discussed in § 2.3.

Similar reasoning applies for the phosphocreatine[1] and inorganic phosphate

---

[1] The phosphocreatine signal that is often present in the ^{31}P spectra of intact tissues pro-vides a suitable and very convenient chemical shift reference. This is because phosphocreatine has a pK_a value of about 4.6, and its resonance frequency is therefore insensitive to pH changes within the normal physiological range. For this reason, and because there are problems associ-ated with using phosphoric acid, which is a commonly used reference (see § 6.1.2), the ^{31}P chemical shift values quoted in this book are, unless stated otherwise, given relative to the signal of phosphocreatine at pH 7, which is assigned the value 0 p.p.m.

signals, both of which can be unambiguously assigned. It is interesting to note that the inorganic phosphate signal of Fig. 2.1 is split into two components that are shifted slightly from each other. The explanation for this is that inorganic phosphate is present within both the intracellular space and the perfusion medium. These two environments have different pH values, and therefore generate inorganic phosphate signals of slightly different frequencies (see § 2.2). In order to confirm that the signals do not arise from two different intracellular metabolites, one could show that the spectra of heart extracts contain only one signal in the inorganic phosphate region.

The signal at −2.5 p.p.m. is characteristic of the γ-phosphate group of ATP, but could also contain a contribution from the β-phosphate group of adenosine disphosphate (ADP). Similarly, the signal at −7.5 p.p.m. could contain contributions from the α-phosphate groups of both ATP and ADP. (In fact, the ATP and ADP signals are shifted slightly from each other, as shown in Fig. 2.2 (b), but in many spectra of living systems they are not sufficiently separated for them to be resolved.) In addition, a shoulder to the right of this peak can often be seen at a chemical shift characteristic of the phosphate signals from NAD[+] and NADH. This shoulder is clearly visible in the kidney spectrum shown in Fig. 2.4. The relative contributions of the various compounds are best determined by accurate comparison of the intensities of the three peaks at −2.5, −7.5, and −16 p.p.m. For example, if the

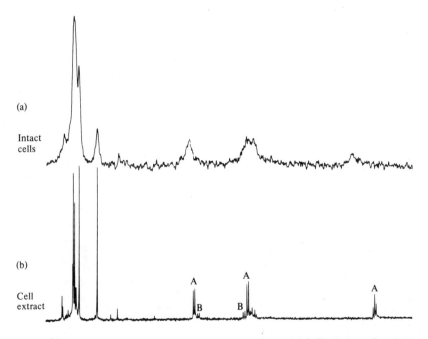

(a)

Intact cells

(b)

Cell extract

A

A

B   B

A

Fig. 2.2. ^{31}P n.m.r. spectra obtained at 145.7 MHz of (a) Ehrlich ascites tumour cells, and (b) a perchloric acid extract of these cells. The peaks labelled A and B are assigned to ATP and ADP respectively. (Adapted from Navon, Ogawa, Shulman, and Yamane, 1977a.)

relative areas of the signals at -2.5 and -16 p.p.m. were 1.3:1, then we could conclude (if the appropriate controls had been performed (see § 6.4)) that the relative concentrations of ATP and ADP were 1:0.3.

^{31}P n.m.r. spectra of cells and tissues frequently contain signals in the region around 7 p.p.m. (measured relative to phosphocreatine). It is often extremely difficult to assign these signals because they could be generated by any of a large number of compounds; any phosphomonoester could produce a signal in this region. Only in a few cases can conclusive assignments be made. For example, a large increase in signal intensity is observed in this region in the spectra of hearts perfused with deoxygluocose (Bailey, Williams, Radda, and Gadian 1981) which is totally consistent with the anticipated formation of deoxyglucose 6-phosphate. Similarly, the formation of fructose 1, 6-diphosphate is detected from the spectra of iodoacetate-poisoned muscles (Dawson, Gadian, and Wilkie 1977*b*). These assignments can then be confirmed by experiments performed on extracts.

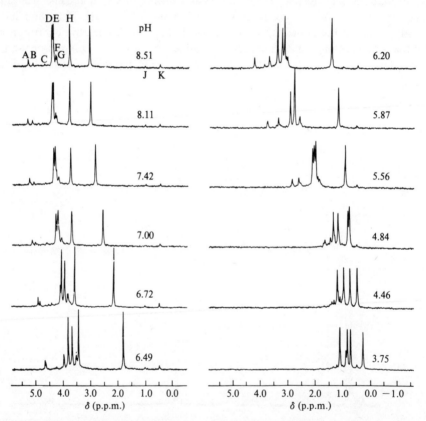

Fig. 2.3. ^{31}P n.m.r. spectra obtained at 145.7 MHz showing the pH dependence of the extract of Fig. 2.2(b). The chemical shifts in this spectrum are expressed relative to phosphoric acid; see footnote on p. 24. (Adapted from Navon *et al.* 1977*a*, where details of the peak assignments are given.)

N.m.r. spectra of extracts provide additional information because the resonances are narrow in comparison with those observed in spectra of intact tissues and cells (see Fig. 2.2). One method of assignment using extracts is to record spectra as a function of pH (see Fig. 2.3). Spectra can be obtained both before and after the addition of expected compounds, and if the additional signal always coincides with the resonance of interest in the extract spectrum this provides further evidence that the assignment is indeed correct. This procedure of adding compounds to the extract is recommended because the chemical shifts of some resonances are sensitive to salt concentration which is difficult to control in extracts. Further confirmation of assignments can be obtained by adding appropriate enzymes to the extract and observing changes in the spectrum that correspond to enzymatic conversion of one compound to another. Alternatively, standard biochemical procedures can be used for identifying compounds within the extracts.

The frequent detection of ^{31}P resonances in the region around 2 p.p.m. (see Fig. 2.4) provides an interesting example of how standard biochemical methods can be combined with n.m.r. observations to assign unidentified resonances. Burt, Glonek, and Bárány (1976a) first reported these unexpected resonances in the 'phosphodiester' region of the spectra from several different types of muscle. Prior to identification of these signals, they were sometimes referred to as 'mystery peaks'. How-

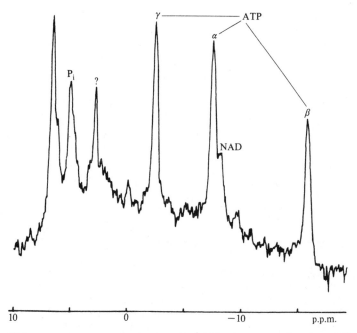

Fig. 2.4. ^{31}P n.m.r. spectrum obtained at 73.8 MHz from a perfused rat kidney. The signal labelled '?' is from a phosphodiester, as yet unidentified. The other signals have similar chemical shifts (though different intensities) to the signals observed in Fig. 2.1, and are assigned similarly; the peak at about 6.5 p.p.m. is from sugar phosphates and AMP/IMP. (From the work of Radda *et al.* 1980.)

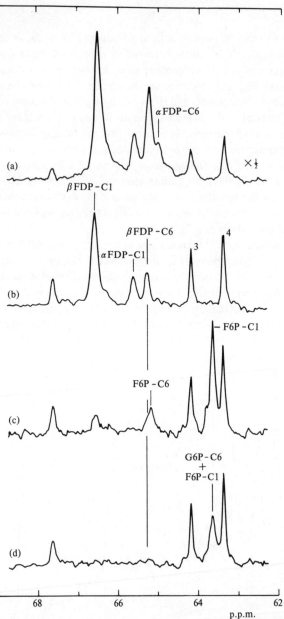

Fig. 2.5. ^{13}C n.m.r. spectra obtained at 90.5 MHz from a perchloric acid extract of *E. coli*. Only the region between 62 and 69 p.p.m. is displayed. (b) is the initial spectrum. (a) shows the effect of adding fructose 1, 6-diphosphate (FDP) labelled at the C-1 and C-6 positions. The resulting spectrum strongly supports the assignments. (c) was the spectrum obtained after the addition of fructose 1, 6-diphosphatase, showing the formation of fructose 6-phosphate (F6P). Spectrum (d) was obtained after subsequent addition of phosphoglucose isomerase, and is consistent with the conversion of fructose 6-phosphate to glucose 6-phosphate. Peak 3 is unidentified, while peak 4 stems from an impurity in the ^{13}C-labelled glucose that was used. (From Ugurbil *et al.* 1978*a*.)

ever, n.m.r. experiments, together with various chemical and chromatographic procedures, confirmed that the major phosphodiester in several muscle types was glycerol 3-phosphorylcholine (Burt, Glonek, and Bárány, 1976*b*; Seeley *et al.* 1976). An additional compound, L-serine-ethanolamine phosphodiester, was identified in a similar manner in the pectoralis muscle of chickens with hereditary muscular dystrophy (Chalovich *et al.* 1977). Several phosphodiester signals have now been observed in the spectra of a variety of living systems, but unambiguous assignment of these signals is not trivial; n.m.r. experiments should always be backed up by additional identification procedures. A search for the possible role of these compounds continues (Chalovich and Bárány 1980).

A final example involves assignments of some of the signals in the ^{13}C n.m.r. spectra of *E. coli*. The assignments of the fructose 1, 6-diphosphate region of the spectrum (see Fig. 2.5) were checked by monitoring the n.m.r. spectra of an extract during enzymatic digestion. Fructose 1, 6-diphosphatase was first added to convert fructose 1, 6-diphosphate to fructose 6-phosphate, and then phosphoglucose isomerase was added to convert the fructose 6-phosphate to glucose 6-phosphate. The resonances behaved exactly as anticipated, verifying that the tentative assignments were indeed correct.

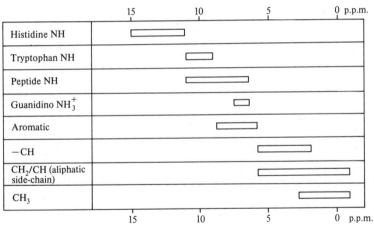

Fig. 2.6. Representative ^{1}H chemical shift values (measured relative to tetramethylsilane (TMS)).

Some chemical shifts for ^{1}H, ^{13}C, and ^{31}P nuclei are given in Figs. 2.6–2.8. Additional aids to the assignment of resonances, which are particularly useful in ^{1}H n.m.r., can be obtained from measurements of coupling constants and from spin-echo and double-resonance experiments (see Chapter 6). The assignment of complex ^{1}H n.m.r. spectra is discussed in depth by Campbell and Dobson (1979).

2.2 pH measurements by n.m.r.

The methods most commonly used for measuring intracellular pH include (i) insertion of a pH-sensitive microelectrode into the cell, (ii) analysis of the distribution of weak acids or bases, and (iii) colorimetry and fluorometry. These

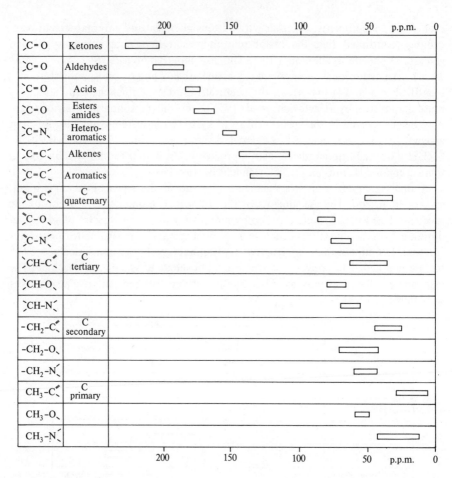

Fig. 2.7. Representative ^{13}C chemical shift values (measured relative to tetra-methylsilane (TMS)).

methods all have advantages and drawbacks that are well documented and have been critically reviewed (Roos and Boron 1981; Cohen and Iles 1975; Gillies and Deamer 1979).

^{31}P n.m.r. provides an alternative method of measuring intracellular pH, as first shown by Moon and Richards (1973). It could be argued that an n.m.r. spectrometer is a most extravagant form of pH meter, but the method does have some extremely important advantages and of course pH is just one of the many parameters that can be measured by n.m.r. The measurement of pH provides an important contribution to our understanding of *in vivo* metabolism. For example, hydrogen ions play a direct part in many enzyme-catalysed reactions, and from this point of view the hydrogen ion can be regarded as an important metabolite. An interesting example is the creatine kinase reaction which is discussed further in §§ 2.5 and 3.5. In addition, the activity of many enzymes is remarkably pH sensitive, and

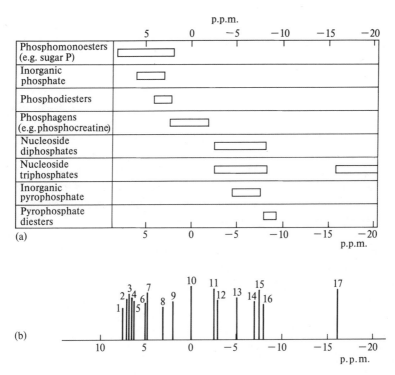

Fig. 2.8. (a) Representative ^{31}P chemical shift values. (The phosphocreatine signal at pH 7 is assigned the value 0 p.p.m.) (b) A 'stick spectrum' illustrating the ^{31}P chemical shifts at pH 7.0 of some compounds commonly found *in vivo*. The phosphocreatine signal is assigned the value 0 p.p.m. 1, dihydroxyacetone phosphate (hydrated); 2, glucose 6-phosphate; 3, glycerol 1-phosphate; 4, fructose 6-phosphate, 5, AMP; 6, glucose 1-phosphate; 7, inorganic phosphate; 8, glycerophosphorylcholine; 9, phosphoenolpyruvate; 10, phosphocreatine; 11, the γ-phosphate of MgATP; 12, the β-phosphate of MgADP; 13, inorganic pyrophosphate; 14, the α-phosphate of MgADP; 15, the α-phosphate of MgATP; 16, NADH; 17, the β-phosphate of MgATP. Phosphoric acid would appear on this scale at anywhere between 2.3 and 3.2 p.p.m., depending only on the sample geometry. For signals that are split by spin-spin coupling, the centres of the coupling patterns are given. Data from Gadian *et al.* (1979) where pH titrations are given for a wide range of phosphorus-containing compounds.

knowledge of intracellular pH therefore helps to provide an understanding of the activities of these enzymes *in vivo*. The pH of many tissues and organs provides an invaluable guide to their metabolic state; this is because a decline in pH tends to reflect the production of lactic acid which accumulates only under ischaemic or stressful conditions. In fact the pH change can sometimes be used to assess the amount of lactic acid formed. The free energy of hydrolysis of ATP *in vivo* can now be calculated (see § 3.6) and this measurement requires a knowledge of the intracellular pH. Mitchell's chemiosmotic theory predicts that pH gradients should be of major importance in a wide range of bioenergetic systems, and the measurement of pH *in vivo* should provide a valuable contribution to our understanding of these systems. Finally, there is much interest in the pH changes that are associated with the activation of metabolic processes in a wide range of organisms, and in the general question of how intracellular pH is regulated.

N.m.r. provides a particularly useful means of monitoring intracellular pH because measurements can be made non-invasively and continuously. In principal, any resonance whose frequency is sensitive to pH can provide an indication of pH, but in practice the resonance of inorganic phosphate is most commonly used because it is readily observable in the majority of ^{31}P spectra and because its frequency is particularly sensitive to pH in the physiological pH range.

Inorganic phosphate exists mainly as HPO_4^{2-} and $H_2PO_4^-$ at around neutral pH. In the absence of chemical exchange these two species would give rise to two resonances separated from each other by about 2.4 p.p.m. In solution, however, the two species exchange with each other very rapidly, and as a result the observed spectrum consists of a single resonance, the frequency of which is determined by the relative amounts of the two species (see § 6.5). The frequency of the signal measured as a function of pH therefore produces the usual type of pH curve (see Fig. 2.9). In principal, it should be possible to determine intracellular pH simply by measuring the chemical shift of the inorganic phosphate resonance *in vivo* and determining from the standard titration curve the pH to which this chemical shift corresponds. However, there are a number of potential problems that must be overcome.

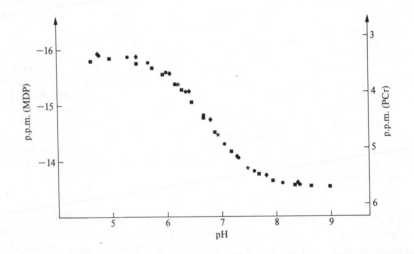

Fig. 2.9. Variation of the chemical shift of inorganic phosphate with solution pH at 37°C. The chemical shift is expressed relative to two different standards; (i) methylene diphosphonate (MDP), and (ii) phosphocreatine (PCr) at pH 7. Solutions containing 10 mM inorganic phosphate and 10 mM phosphocreatine were adjusted to various ionic strengths by addition of KCl or NaCl, and titrated at 37°C by addition of HCl and KOH or NaOH. Symbols: ◆ 120 mM KCl; ■ 160 mM KCl; ● 200 mM KCl; ★ 160 mM NaCl. (Adapted from Garlick, Radda, and Seeley 1979.)

First, we must consider whether any factors other than pH can affect the chemical shift of inorganic phosphate *in vivo*, for if such effects exist they could generate errors in the estimates of intracellular pH. Certainly, large variations in ionic strength (Gadian, Radda, Richards, and Seeley 1979), metal ion binding (Jacobson and Cohen 1981), and temperature can influence the chemical shift of inorganic phosphate, largely through their effects on the pK_a of this compound. Therefore, it is always advisable to perform the standard titration curve using a solution whose ionic composition resembles that found *in vivo*. Of course, there is bound to be some uncertainty regarding the precise ionic environment within the cell, but fortunately control experiments (such as those shown in Fig. 2.9) have shown that the errors generated by these uncertainties are likely to be very small.

The binding of inorganic phosphate to macromolecules is another factor that could influence its chemical shift *in vivo*. However, it has been found that inorganic phosphate titration curves are similar in simple aqueous solution and in homogenized dog heart preparations (see Hollis 1979). This suggests that phosphate–protein binding interactions have little effect on the observed inorganic phosphate chemical shifts. Additional evidence that such binding has little effect on the chemical shift is essentially circumstantial; if there were differences between intracellular pH measured in this way and pH as measured by other means, one might suspect that the discrepancy might be explained on the basis of binding effects. However, where comparison is possible the agreement is good (see below), and it is therefore reasonable to conclude that any effects of binding are small and can be ignored.

An additional problem is that the chemical shift of any resonance is measured by comparing the frequency of that resonance with that of a reference compound. It is therefore essential to have available a suitable reference relative to which chemical shifts can be determined. For [31]P n.m.r., this is provided by phosphocreatine when present. Alternatively, the ^{1}H signal from the water within biological samples provides an acceptable frequency standard and will probably become increasingly used in the future (Ackerman, Gadian, Radda, and Wong 1981). The use of standards is discussed in detail in § 6.1.2.

A final problem is that there might be some uncertainty as to the distribution of inorganic phosphate within the various intracellular compartments. For example, mitochondria occupy about 40 percent of the intracellular volume of the heart, and if the majority of the inorganic phosphate were located within the mitochondria we might inadvertently be measuring mitochondrial rather than cytoplasmic pH. However, experiments utilizing deoxyglucose (see below and Chapter 3) indicate that in the rat heart [31]P n.m.r. does indeed monitor cytoplasmic pH.

Because of these potential difficulties, it is essential to confirm the reliability of pH measurements using [31]P n.m.r. In fact, there is now a considerable amount of experimental evidence confirming that [31]P n.m.r. provides an acceptable method of measuring intracellular pH. In general, pH measurements made by [31]P n.m.r. in a variety of samples under a variety of conditions agree well with anticipated values. A number of more specific experiments have been performed that verify the reliability of the method. For example, good agreement between [31]P n.m.r. and microelectrode measurements has been found by Nuccitelli, Webb, Lagier, and

Matson (1981) in their interesting studies of *Xenopus laevis* eggs. Using ^{31}P n.m.r. they measured the average intracellular pH of unfertilized, fertilized, and activated eggs to be 7.42, 7.66, and 7.64 respectively. These values are almost identical to the values that were obtained using pH-sensitive glass microelectrodes. Agreement between the two methods was also obtained in studies of the giant barnacle muscle cell (Bagshaw, Gadian, Radda, and Vaughan-Jones, unpublished observations). The pH of these cells was measured to be 7.35 by ^{31}P n.m.r., while the pH of similar cells from the same barnacle was measured to be 7.25 using a microelectrode. Typical values in the literature for the intracellular pH under similar conditions are around 7.30. Furthermore, in experiments where two different resonances provide independent measures of pH (for example, in studies on hearts perfused with 2-deoxyglucose for which both the inorganic phosphate resonance and the 2-deoxy-glucose 6-phosphate resonance provide estimates of intracellular pH (see Chapter 3)) the values obtained using the two resonances are very similar. Finally, the pH of heart and skeletal muscle can be measured using either phosphocreatine or water as the reference compound, and the two references give similar pH values (Ackerman *et al*. 1981). For all of these reasons it is clear that ^{31}P n.m.r. provides a convenient and reliable means of estimating pH. Nevertheless, in view of the uncertainties inherent in the method (including the fact that different pH titration curves that have been reported in the literature differ slightly from each other[1]), it is unwise to hope for *absolute* accuracy of better than 0.1 pH unit when measuring intracellular pH in this way. However, *changes* in pH can often be measured to better than 0.05 pH unit.

In certain circumstances intracellular pH can also be deduced by ^{1}H n.m.r., for the ^{1}H chemical shifts of histidine residues are sensitive to pH. The pH dependence of the histidine resonances of haemoglobin have been carefully characterized by Brown and Campbell (1976). It is therefore possible to determine the intracellular pH of red blood cells from observations of their haemoglobin resonances, as described by Brown, Campbell, Kuchel, and Rabenstein (1977).

It is interesting to note that n.m.r. can provide information about heterogeneity of pH. For example, the inorganic phosphate signal is sometimes split into two distinct components, implying that the sample contains two environments of differing pH (see Fig. 2.1 and § 3.2; see also Busby *et al*. 1978). Alternatively, if there is a range of environments within a sample of differing pH values, the sample will produce a range of inorganic phosphate signals that are very slightly shifted from each other, the net effect being that the observed signal is broadened. In many ^{31}P spectra of skeletal muscle, the inorganic phosphate signal is significantly broader than the signal from phosphocreatine, and this has been interpreted in terms of a distribution of pH within the muscle (Seeley *et al*. 1976; see also Fig. 4.4).

---

[1] Illingworth (1981) has recently pointed out that surprisingly large errors can occur when measuring pH with combined glass electrodes. The pH titration curves that are used for n.m.r. studies are obtained by measuring chemical shifts as a function of solution pH, which is commonly determined with a combined glass electrode. Therefore the errors associated with these electrodes could account for the differences that have been observed between different pH titration curves.

2.3 Metal ion binding

The most important study of this type concerns the interaction of Mg^{2+} ions with ATP. The pioneering work in this area was performed by Cohn and Hughes (1962) who showed that the three ^{31}P resonances of ATP are considerably shifted on binding of metal ions such as Mg^{2+}. Unfortunately, their results are not immediately relevant to studies of living systems because spectrometer sensitivity in those days was such that they had to use 100 mM ATP. At this high concentration there is considerable stacking of the aromatic rings of the ATP molecules which can affect the resonance frequencies. pH titrations of 5 mM ATP in the presence and absence of Mg^{2+} ions are shown in Fig. 2.10. It is clear that the frequencies of the ATP

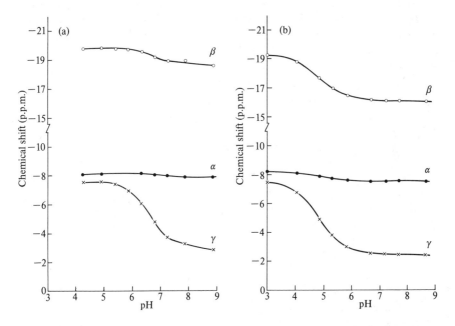

Fig. 2.10. The chemical shifts of the ^{31}P n.m.r. signals of 5 mM ATP plotted as a function of pH. (a) in the absence of Mg^{2+} ions, and (b) in the presence of 10 mM $MgCl_2$, when the ATP is almost fully complexed to Mg^{2+}. (From the data presented in Gadian *et al.* 1979.)

resonances observed *in vivo* should provide information about the extent to which ATP is complexed to Mg^{2+} ions in the cell. (In principle, binding could also occur with other metal ions such as Ca^{2+}, but in general Mg^{2+} is the only ion present at high enough concentrations to allow significant binding.) The general conclusion that can be reached is that in all biological tissues that have been investigated by n.m.r. the ATP is predominantly complexed to Mg^{2+} ions. More specifically, from their n.m.r. studies on frog skeletal muscle, Gupta and Moore (1980) suggest that about 93 per cent of the ATP is complexed to Mg^{2+} ions. From a knowledge of the binding constant of Mg^{2+} to ATP, it has been concluded that the concentration of free Mg^{2+} ions in muscle cells is about 0.6 mM. However, the figure of 93 per cent relies on very precise measurements of chemical shifts and also on the assumption that no intracellular features other than Mg^{2+}, pH, and ionic strength have any detectable effect on the ATP chemical shifts. While this may well be true, it remains to be proved, and indeed it would be surprising in view of the very specific way in which MgATP binds to a variety of macromolecules. It seems safest to state that the observed chemical shifts indicate a lower limit of about 0.5 mM for the free Mg^{2+} concentration. It is difficult to give an accurate upper limit because the observed chemical shifts are not totally incompatible with, for example, 97 per cent of the ATP being complex to Mg^{2+} rather than 93 per cent. If 97 per cent were the correct figure, this would lead to a value of about 1.5 mM for the concentration of free Mg^{2+} ions. Nevertheless, it is gratifying that the value obtained for free Mg^{2+} is in broad agreement with the value of about 1 mM which has recently been used by other workers (Veech, Lawson, Cornell, and Krebs 1979).

In another rather more complicated series of experiments, Cohen and Burt (1977) studied the interaction of Mg^{2+} ions with phosphocreatine and concluded that the concentration of free Mg^{2+} in muscle cells is about 3 mM. However, this figure was obtained using assumptions that may not be totally valid (see Gupta and Moore 1980).

Measurements of the binding of Mg^{2+} to ATP are of importance because it is MgATP rather than free ATP which is the substrate for most reactions involving ATP. In addition, knowledge of the concentration of free Mg^{2+} is invaluable when studying enzymes such as creatine kinase and adenylate kinase for which the equilibrium constant is sensitive to Mg^{2+} concentration (Lawson and Veech 1979). Moreover, many enzymes require Mg^{2+} ions for activity, and therefore their activity could be strongly dependent on the free Mg^{2+} concentration. These n.m.r. measurements are therefore most welcome, particularly in view of the difficulties associated with measuring the intracellular distribution of Mg^{2+} by other means.

If paramagnetic centres, such as transition metal or lanthanide ions, are present within a sample, they can greatly affect the frequencies and/or relaxation times of n.m.r. signals, and because these effects depend strongly on interatomic distance such centres are frequently used as probes of molecular structure. For studies of living systems it is important to note that trace quantities of paramagnetic ions can produce particularly powerful relaxation of ^{31}P resonances because the ions bind to the negatively charged phosphates. For example, from the results of Brown *et al.* (1973) it can be concluded that the addition of $1 \mu M Mn^{2+}$ to a solution containing 5 mM ATP would broaden the ^{31}P resonances of ATP by about 10 Hz at room temperature. It is therefore essential to remove or chelate paramagnetic ions for many ^{31}P n.m.r. studies of solutions. This can be achieved most easily by adding chelating agents such as ethylene diamine tetraacetic acid (EDTA), and is a procedure particularly recommended for perchloric acid extracts which often contain large amounts of paramagnetic impurities. EDTA binds Mg^{2+} ions much less tightly than other divalent metal ions, and it is therefore a most useful chelating agent for studies in which Mg^{2+} ions are required.

2.4 Concentration and kinetic measurements

The relative concentrations of molecules within a sample can be determined from the relative areas of their resonances, *provided that* the appropriate controls have been performed (see § 6.4). Therefore, simply by monitoring resonance areas as a function of time, the kinetics of reactions can be followed as they proceed within living systems. This is one of the most useful and informative aspects of n.m.r. and a number of applications are discussed in Chapter 3.

The time resolution of kinetic measurements is limited by the time necessary to obtain a satisfactory n.m.r. spectrum. This depends, of course, on many factors, including the concentrations of the compounds of interest, the signal-to-noise ratios that are required, and the type of spectrometer that is used, but for most experiments it can be anticipated that time resolution will be of the order of 1–5 min. In situations where rapid and repetitive changes occur, time resolution can be enhanced by synchronizing n.m.r. data collection with different phases of the cycle. For example, electrical stimulation of frog sartorius muscles has been used to gate ^{31}P n.m.r. measurements of the metabolic events associated with contraction (Dawson, Gadian, and Wilkie, 1977a). In addition, experiments on the fluctuations of phosphocreatine and ATP concentrations during the contraction–relaxation cycle of perfused hearts have been reported by Fossel, Morgan, and Ingwall (1980).

It is important to note that measurements of concentration are not restricted just to those compounds that produce detectable signals. For example, we have already seen that the concentration of H^+ or Mg^{2+} ions in solution can be deduced indirectly from the chemical shifts of some of the resonances. In addition, further important deductions can often be made by combining n.m.r. intensity measurements with information obtained by other means. For example, a knowledge of the equilibrium constant for the creatine kinase reaction coupled with the knowledge that this reaction is close to equilibrium in resting skeletal muscle permits a measurement to be made of the concentration of free ADP. Similarly, it may be possible to use the adenylate kinase reaction to estimate the concentration of free adenosine monophosphate (AMP). The determination of these levels should provide a most important contribution to our understanding of metabolism and its control (see § 3.1).

2.5 Steady state kinetics – exchange processes

N.m.r. spectra are sensitive to chemical exchange processes that take place under steady state or equilibrium conditions (see § 6.5). Of particular interest for studies of enzyme-catalysed reactions *in vivo* are two different types of 'magnetic-labelling' experiment.

The first type of experiment employs an n.m.r. technique known as saturation transfer, the theory of which is discussed in Appendix 6.1. The nature of an experiment designed to investigate the activity of creatine kinase in frog muscle is illustrated in Fig. 2.11. Figure 2.11 (a) shows a normal spectrum of frog gastrocnemius muscles. In the spectrum of Fig. 2.11 (b) a selective radiofrequency pulse is applied to the γ-phosphate peak of ATP in such a way as to cause it to disappear as a result of saturation. (Effectively, the γ-phosphate is being labelled with zero magnet-

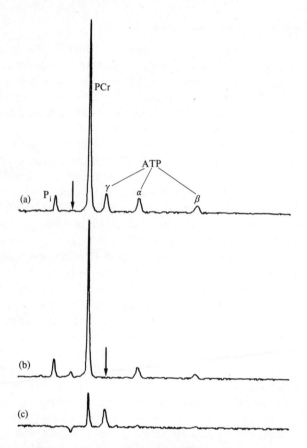

Fig. 2.11. ^{31}P n.m.r. spectra of frog gastrocnemius muscles obtained at 73.8 MHz, with selective irradiation applied at the frequencies indicated by the arrows. The irradiation applied in (a) acts as the control for (b), in which selective irradiation is applied to the signal from the γ-phosphate of ATP. The difference between (a) and (b), given in (c), gives the extent of saturation transfer from the γ-phosphate of ATP to phosphocreatine. (Adapted from Gadian *et al.* 1981.)

ization.) If the γ-phosphate is exchanging with the phosphorus of phosphocreatine, then the *lack* of signal intensity will be transferred from ATP to phosphocreatine, resulting in a reduction in the intensity of the phosphocreatine signal. The extent of this reduction (see Fig. 2.11 (c)) together with measurements of the spin-lattice relaxation time T_1 permits the rate of interconversion to be evaluated. This particular exchange process takes place by means of the reaction catalysed by creatine kinase, and saturation transfer experiments have provided some very interesting results about the activity of this enzyme *in vivo* (see § 3.5).

Unfortunately, only a limited number of enzyme-catalysed reactions can be measured in this way, firstly because they need to be fairly rapid and secondly because they must involve at least one substrate that gives rise to a detectable signal. However, studies of ATPases, and of the important enzyme adenylate kinase are feasible, and it should be stressed that this technique does provide for the first time a totally non-invasive means of measuring the steady state kinetics of certain enzyme-catalysed reactions as they take place within living cells and tissues.

In the second type of experiment a magnetic label in the form of a suitable isotope is present within the sample and its effect on the spectra is monitored. For example, if red blood cells are suspended in a ^2H_2O solvent, the ^1H n.m.r. signals from the C-3 protons of added lactate and pyruvate gradually disappear because the C-3 position becomes deuterated. The rate of interconversion of lactate and pyruvate can be measured from the observed exchange rates, and hence the activity of lactate dehydrogenase within the cells can be determined (see § 3.4). Exchange at the C-2 position of lactate is more complex, but it provides information about the rates of additional reactions of the glycolytic pathway.

In experiments of a similar type ^{13}C n.m.r. can be used to follow the metabolism of a ^{13}C label introduced into a living system. The observed distribution of the ^{13}C label among the carbons of intermediates and products can provide detailed information about the flux rates through some of the reactions (see § 3.3).

2.6 Compartmentation

If we are to obtain a thorough understanding of cellular metabolism, it is essential to know how the metabolites are distributed within the cell. Unfortunately, however, we have very limited knowledge about the 'compartmentation' of metabolites within cells. Compartmentation does not refer only to the localization of compounds within membrane-bound compartments of a cell; in a more general sense it refers to the existence of different pools of a given compound that exchange very slowly with each other. For example, the ATP:ADP ratio is very different in the cytoplasm from its value within mitochondria; this provides a fairly obvious and important example of compartmentation. However, a less obvious example is provided by the pool of ADP that is tightly bound to the myofilamentary proteins of skeletal muscle. This pool of ADP is inaccessible to most of the soluble enzymes and can be considered to be compartmented away from the ADP that is freely mobile within the cytoplasm. It should be noted that these two pools of ADP are not separated by any membrane.

N.m.r. can provide information about intracellular compartmentation, provided that the different pools of any given compound give rise to recognizably different n.m.r. signals, and a number of examples are given in § 3.2.

N.m.r. can also be used to distinguish between intracellular and extracellular pools of metabolites. For example, transport of metabolites across cell membranes can sometimes be studied (see § 2.9), and it is possible to monitor pH gradients across the cell membranes of systems such as *E. coli* (see § 3.2). In addition, as a general procedure internal and external pools of metabolites can be distinguished by adding paramagnetic ions to broaden or shift the signals from the external medium. Of course, this type of experiment relies on the ions not penetrating into the internal medium.

2.7 Oxidation state

The ^{1}H n.m.r. spectrum of glutathione depends on its oxidation state as shown by Brown *et al.* (1977*a*) in their studies of red cells. ^{1}H n.m.r. can therefore provide a valuable non-invasive measure of the oxidation state of these cells, and perhaps of other systems that contain significant quantities of glutathione.

Optical spectroscopy is particularly well suited for studying the redox state of living systems (Sies and Brauser 1980), and there would be much to be gained by making simultaneous observations with n.m.r. and with fluorescence or absorption spectroscopy. Relatively simple modifications to instrumentation should permit this type of simultaneous measurement to be feasible.

2.8 Molecular mobility

N.m.r. spectra can provide information about many different types of molecular motion. This section relates to overall molecular rotation or tumbling, and also to internal motion or flexibility within molecules. Other types of motion that involve translation of molecules through space are discussed in § 2.9.

Basically, there are two ways in which n.m.r. can be used to detect mobility. One relies on exchange processes which are discussed in § 6.5, and the other relies on observations of relaxation times, linewidths, and the nuclear Overhauser effect (see Chapter 6). These two approaches, and their applications to proteins and membranes, have been reviewed in some detail by Campbell and Dobson (1979) and by Campbell (1977). Unfortunately, the depth of analysis that is used for protein studies is not really feasible when studying living systems for which there are too many uncertainties about the nature and effectiveness of the various types of relaxation and exchange processes. As a result we have to be content with a very qualitative idea of the extent of molecular motion within living systems.

An obvious example of the way in which n.m.r. spectra respond to mobility differences is given by the observation that membrane phospholipids give rise to much broader n.m.r. signals than metabolites. However, a number of more interesting examples can be given, all of which rely on the fact that as molecules become more immobilized, their resonance linewidths tend to increase. No ADP signal is detectable in the ^{31}P n.m.r. spectra of skeletal muscle, although the total concentration is estimated from freeze-clamping studies to be about 0.5 mM. The absence of signal is consistent with the belief that much of the ADP is tightly bound to the proteins of the myofilaments; presumably the ADP is so immobilized that its signals are too broad to detect. However, it should be pointed out that, although the signal of a metabolite will broaden on binding to a cytoplasmic enzyme, the broadening will not in general be so great as to render the signal undetectable. This is because the bound metabolite can still retain a fair degree of mobility, either through internal motion or via the overall mobility of the enzyme itself. Evidence for this comes from the studies of Cohn and coworkers on enzyme solutions. They observed fairly narrow ^{31}P n.m.r. signals from metabolites bound to a number of enzymes including arginine kinase, creatine kinase, and adenylate kinase (see Cohn and Rao (1979) for a review).

As another example of the effects of immobilization, the membrane-bound granules within human blood platelets contain large quantities of ATP and yet produce no high-resolution ^{31}P n.m.r. signals. This is presumably because the mobility of the ATP within the granules is highly restricted (Ugurbil, Holmsen, and Shulman 1979). If we now turn to ^{1}H n.m.r., the ^{1}H spectra of red blood cells contain narrow signals from metabolites superimposed on relatively broad signals from the haemoglobin. It is possible to use a spin—echo sequence of n.m.r. pulses (see § 6.3.7) to eliminate the broad signals, leaving a much simplified spectrum that consists predominantly of signals from the metabolites. Here, the difference in linewidths is used as a means of spectral simplification rather than as a means of assessing the rates of molecular motion. As another example, many of the spin-imaging techniques generate images that depend not only on proton density but also on the proton relaxation times. It is therefore possible to discriminate between regions that contain protons of differing degrees of mobility. For example, bone marrow, which contains mobile protons, generally produces much greater signals than the cortex of bone in which there is a very low density of mobile protons.

Finally, it should be noted that sophisticated n.m.r. techniques now enable high-resolution signals to be observed from highly immobilized structures such as membranes and bone (Herzfeld *et al.* 1980). These techniques of solid state n.m.r. are outside the scope of this book, but there are many exciting possibilities, exemplified by some particularly interesting ^{15}N studies of ^{15}N labelled soybeans (Schaefer, Stejskal, and McKay 1979).

2.9 Imaging, diffusion, flow, and transport

We shall see in Chapter 4 that the use of well-defined magnetic field gradients permits 'images' to be obtained that represent the spatial distribution of molecules within a sample. Field gradients can also be used in conjunction with pulsed n.m.r. techniques to study the movement of molecules via diffusion (see § 6.3.7), flow, or transport. Many of these studies are performed on water, not just because the large concentration of water facilitates measurement but also because of the intrinsic value of measuring parameters such as blood flow.

Molecular diffusion measurements using n.m.r. are usually made over times ranging up to 100 ms. It is therefore possible to estimate the true intracellular diffusion constants, unaffected by membrane barriers, because the diffusion path lengths within these short periods are smaller than the linear dimensions of large cells. Studies of water in tissues such as muscle show that the translational mobility of most water molecules in the cytoplasm is at least half that of ordinary water (Finch 1979; Shporer and Civan 1977). On the basis of these particular n.m.r. measurements, it is unnecessary to endow the bulk of the intracellular water with any particularly unusual properties. Nevertheless, a great deal remains to be understood about the properties of intracellular water.

There is an extensive literature relating to n.m.r. measurements of flow (see Jones and Child 1976). For biological studies the most important application is undoubtedly the study of blood flow, and the possibilities were recognized many years ago (see for example Singer 1959). However, there seem to be surprisingly few reports of blood flow measurements in animals or human beings (for examples see Morse and Singer 1970; Battocletti *et al.* 1979). One interesting type of experiment relates to the observation that the presence of deoxyhaemoglobin, which is paramagnetic, decreases the observed relaxation time T_2 of the blood water protons (Thulborn, Waterton, Styles, and Radda 1981). It may therefore be possible to assess the oxygenation state of the haemoglobin within arterial and venous blood vessels by measuring the blood water spin–spin relaxation time. Such measurements together with blood flow determinations could enable estimates to be obtained for the oxygen consumption of organs within animals.

Transport of metabolites across cell membranes can be studied if internal and external pools of metabolites can be distinguished from each other. For example, Brindle *et al.* (1979) have shown that in the ^1H spectra of red cell suspensions the signals from internal alanine are of much greater intensity than those from alanine external to the cells. They conclude that this is because the presence of densely packed cells generates magnetic field inhomogeneities that are much greater outside the cells than inside. After adding alanine to the suspending medium, there is a gradual growth of signal which provides a direct measure of the rate of transport of alanine into the red cells.

3

Applications to cells and tissues

In this chapter, we present a few examples of how n.m.r. can be used to study some specific problems that confront the biologist. The aim is *not* to provide a review of the field; this would merely duplicate or extend the many reviews that have recently appeared in the literature. Many of the metabolic studies that we shall discuss relate to the glycolytic pathway (see Fig. 3.1), and before proceeding to a description of these studies, it may be instructive to discuss briefly the type of information that we would like to obtain about this, or any other pathway.

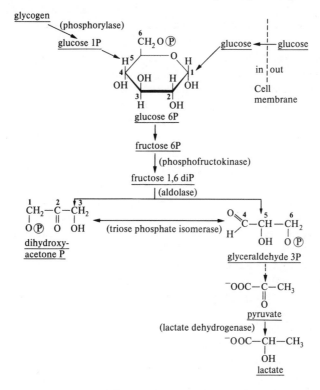

Fig. 3.1. Selected steps in the glycolytic pathway.

Essentially, we would like to measure the rates at which the various reactions of the pathway take place *in vivo*, and we would also like to understand how these rates are controlled in such a way that metabolic fuel is provided according to demand. For example, how is it that the rate of lactate production in muscle can be enhanced by a factor of up to 1000 following the initiation of muscular contraction? Theories of metabolic control have certainly been proposed that could account for such an enhancement of metabolic activity (see, for example, Newsholme and Start (1973)), but they are at least partially based on studies performed *in vitro*. It is therefore important to establish the validity or otherwise of these theories, by measuring metabolic activity in a fully functioning preparation. Moreover, it is essential to measure the concentrations *in vivo* of metabolites such as ATP, ADP, and AMP that are believed to play key roles in regulating the activity of enzymes such as phosphofructokinase. If we are to understand the roles of these metabolites, it is particularly important to relate *changes* in their levels to *changes* in metabolic activity. As we shall see, this type of study is now feasible using NMR.

Similarly, we would like to measure the concentrations of compounds such as NAD^+ and NADH that are involved in cellular redox reactions. Optical spectroscopy is particularly well suited for studying the redox state of living systems, and for this reason it should be most interesting to combine high resolution n.m.r. studies with simultaneous measurements of redox state using fluorescence or absorption spectroscopy. However, all of these measurements of metabolite levels only have real value if we can define the intracellular environment of the various compounds, and that includes identifying the intracellular compartments within which the compounds are located. In addition, we must measure the rates at which the various substrates and products are transported across the cell membrane and between intracellular compartments. A knowledge of these rates is essential if we are to thoroughly understand the functioning of any metabolic pathway.

It is also of interest to establish which reactions in a pathway are close to equilibrium, and which are far removed from equilibrium. The reason for this is that the role of an enzyme in the organisation of a metabolic pathway can be classified according to whether the reaction is close to, or far removed from equilibrium (see Newsholme and Start 1973). For example, the net flux through a pathway is determined by the rates of the non-equilibrium reactions. Although one can predict, on the basis of *in vitro* studies, the class to which each enzyme within a pathway belongs, it would certainly be useful to establish by measurements performed *in vivo* whether these predictions are correct.

Finally, little is known about the relationships between metabolism and physiological activity, largely because of the difficulties involved in studying the metabolism of a fully functioning system. N.m.r. experiments open up the possibility of measuring metabolite levels, fluxes thrugh metabolic pathways, rates of reactions taking place under steady state or equilibrium conditions, and rates of transport; in addition, one can monitor intracellular environment and compartmentation. All of these measurements can be made on a fully functioning system, and therefore there is every reason to believe that such studies should enhance our understanding of how metabolic and physiological processes are controlled and integrated.

3.1. Concentrations of phosphorus-containing metabolites

For a variety of reasons it is important to compare the absolute or relative concentrations of metabolites as measured by n.m.r. with those measured by freeze extraction. Agreement, in cases where agreement is expected, should confirm the validity of the two approaches. Alternatively, disagreement might cast doubt on the accuracy of one or other of the approaches, or, more interestingly, could provide information about the nature or environment of the metabolites within the cell. A number of interesting comparisons are beginning to emerge in the case of the phosphorus-containing metabolites.

In comparing concentrations measured by the freeze-clamping technique with those measured by n.m.r. it is important to note that freeze-clamping estimates usually rely on measurements of *total* amounts of metabolites, whereas n.m.r. generally measures only the mobile components. Therefore, the values obtained by the two techniques could differ as a result, for example, of tight binding of metabolites to macromolecules within the cell.

For frog sartorius muscle, freeze-clamping and n.m.r. provide similar values for the concentrations of ATP, phosphocreatine, and inorganic phosphate (Dawson *et al.* 1977*a*), which indicates that only a small fraction of any of these compounds can be substantially immobilized. These muscles are thin and therefore can be rapidly frozen. For this reason, there are no problems (see below) associated with the breakdown of high-energy phosphates during the process of freeze-clamping.

In contrast, a wide range of studies on various tissues (including the studies of live anaesthetized rats that are described in §§ 1.9.2 and 4.1.1) show that the analytically measured ADP concentration is generally much greater than that estimated by n.m.r. Moreover, in skeletal muscle it is possible to obtain an accurate measure of the free ADP concentration from the creatine kinase equilibrium, and the free ADP estimated in this way is only a few per cent of the total ADP. The same conclusion regarding free ADP has been reached by Veech *et al.* (1979) on the basis of similar arguments involving enzymes such as creatine kinase that are believed to catalyse reactions that are near to equilibrium in the cell; they suggest that free ADP is very low in brain, muscle, and probably liver. Before discussing the important implications of these observations, it is worth commenting on the possible reasons for this discrepancy between n.m.r. and freeze-clamping measurements.

There is a straightforward explanation for the low n.m.r. value for free ADP in muscle, for it has been assumed for some time that a large percentage of the ADP is tightly bound to the proteins of the myofilaments (see Veech *et al.* 1979 for references). Presumably, this ADP is too immobilized to generate detectable n.m.r. signals, as first suggested by Bárány *et al.* (1975). However, this explanation does not account for the discrepancy observed in other tissues. An alternative possible explanation for the disagreement between n.m.r. and freeze clamping is that for many tissues there is unavoidable breakdown of high-energy phosphates during the process of freeze clamping and subsequent extraction. Not only would this account for these ADP results, but it would also explain the additional observation that the inorganic phosphate levels determined by n.m.r. are usually very much lower than those measured in freeze-clamping studies. A further possibility is that a significant fraction of the intracellular content of compounds such as ADP could be sequestered, e.g. in the mitochondria, in such a way that it generates no detectable signal.

The fact remains that n.m.r. experiments demonstrate that the concentrations of free ADP and inorganic phosphate are often very low in well-oxygenated tissues and organs. For example, in resting anaerobic frog muscle at 4 °C the concentration of free cytoplasmic ADP is only about 20 μM (see § 3.5), and on the basis of other n.m.r. studies one would anticipate that this value is not atypical of muscle in general. The concentration of inorganic phosphate measured by n.m.r. is about 1–1.5 mM in a variety of different types of muscle (although it seems rather higher in human skeletal muscle; see § 4.1.3) and is almost certainly no greater than this in liver, kidney, and brain.

There are some important thermodynamic implications associated with these observations. In particular, the free energy of hydrolysis of ATP, given by

$$\Delta G = \Delta G^0 + RT \ln \left\{ \frac{[ADP]\,[P_i]\,[H^+]}{[ATP]} \right\},$$

will be considerably greater than might have been anticipated. Equivalently, the so-called phosphorylation potential $[ATP]/[ADP]\,[P_i]$ will be very much higher than the value predicted from freeze-clamping studies. The precise value of the free energy of ATP hydrolysis *in vivo* is of great interest in relation to understanding the thermodynamics of oxidative phosphorylation and can also assist our understanding of active transport processes (see, for example, § 3.6) which rely on the energy derived from ATP hydrolysis.

There are also important kinetic implications of the low values that have been obtained for the concentration of free ADP. In particular, the Michaelis constant K_M for ADP stimulation of respiration is of the order of $20\,\mu M$ (Chance and Williams 1956; Jobsis and Duffield 1967), and therefore it is not surprising to find that the concentration of free ADP is in this region. In addition, if as now seems certain the free ADP concentration is considerably lower than is often quoted, this would profoundly affect our understanding of a number of glycolytic reactions that involve ADP. One example is the reaction catalysed by pyruvate kinase which uses ADP as a substrate.

It is particularly important to note that the concentrations of AMP, ADP, and ATP are linked by the adenylate kinase reaction:

$$AMP + ATP \rightleftharpoons 2\,ADP.$$

If this reaction is at equilibrium, we can write

$$K_{eq} = \frac{[ADP]^2}{[AMP]\,[ATP]}.$$

The equilibrium constant is approximately equal to unity, and it can readily be shown that, if the concentrations of ATP and ADP are 4 mM and $20\,\mu M$ respectively the concentration of free AMP must be only about $0.1\,\mu M$. This value is about two orders of magnitude lower than the value normally quoted for total AMP. AMP is believed to be a most powerful regulator of enzymes such as phosphorylase and phosphofructokinase (see for example Newsholme and Start 1973), and so it is essential to establish whether the adenylate kinase reaction is indeed at equilibrium. If it is, then we must reconsider the various theories of metabolic control in the light of the low levels of free ADP and AMP.

The above discussion is subject to the criticism that there is some uncertainty as to the precise intracellular *location* of the various metabolites. This is a valid criticism, and this difficult problem of compartmentation is discussed in the following section. However, in the absence of substantial evidence to the contrary, it seems reasonable to assume, particularly in white skeletal muscle containing few mitochondria, that the predominant contribution to the observed signals is from metabolites within the cytoplasm. The arguments presented above can therefore be considered to relate to the cytoplasmic compartment. One piece of supporting evidence comes from studies of hearts perfused with 2-deoxyglucose (Bailey *et al.* 1981). In these experiments 2-deoxyglucose 6-phosphate accumulates in the cytoplasm, and its ^{31}P n.m.r. signal provides a measure of cytoplasmic pH. The pH determined from this signal agrees well with the value obtained from the inorganic phosphate signal, confirming that the inorganic phosphate reflects the cytoplasmic rather than the mitochondrial pH.

A logical and important extension of the studies discussed in this section is to monitor *changes* in metabolite levels, and to relate these to changes in the metabolic activity of a fully functioning tissue. Such experiments should considerably enhance our understanding of how metabolic processes are controlled *in vivo*. However, a comprehensive picture will only be obtained by combining n.m.r. studies with additional complementary techniques, such as freeze-clamping, optics, and measurements of oxygen consumption.

3.2. Compartmentation and pH gradients

Lack of knowledge about the precise distribution of metabolites within the cell represents a major gap in our understanding of cellular processes. N.m.r. can at least begin to provide a glimpse of how certain metabolites might be compartmented within the cell, for under certain conditions the n.m.r. signals from given nuclei can be identified with specific intracellular compartments. Information about compartmentation is also emerging from other techniques (see Akerboom, van der Meer, and Tager 1979), and it is to be hoped that significant progress can be made by combining the information available from all the various approaches.

A particularly important question is whether n.m.r. can provide information about the distribution of metabolites between cytoplasm and mitochondria. It is fair to say that studies of whole tissues have so far provided little, if any, conclusive information on this matter; Nunnally and Hollis (1979) have interpreted some of their data on hearts in terms of compartmentation of ATP between the cytoplasm and mitochondria, but alternative explanations can be put forward for their observations (see § 3.5). In view of this dearth of information, it would seem that the most logical procedure is to characterize fully the signals generated by mitochondria. Unfortunately, however, relatively few studies have yet been performed.

Ogawa *et al.* (1978) have obtained ^{31}P spectra from suspensions of rat liver mitochondria at 0 °C (see Fig. 3.2.). After oxygenation with succinate and hydrogen peroxide, they observed separable signals from two pools of ATP and possibly ADP. The signals were assigned to intra- and extra-mitochondrial ATP and ADP, and are resolved because the internal ATP and ADP is predominantly bound to metal ions while the external pool is largely uncomplexed; divalent metal ions produce significant shifts in the signals of ADP and ATP, as discussed in § 2.3. The signals assigned

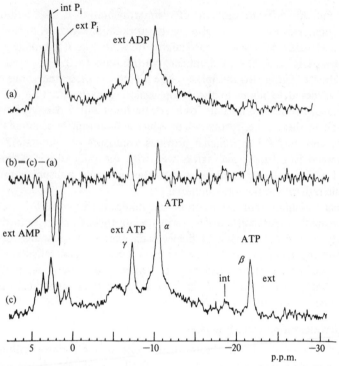

Fig. 3.2. ^{31}P n.m.r. spectra obtained from mitochondria in 10 minutes at 145.7 MHz, at a temperature of 0 °C. (a) Prior to oxygenation; (b) the difference between (c) and (a); (c) after oxygenation with 5 mM succinate and H_2O_2. Chemical shifts here are expressed relative to phosphoric acid. Signals are assigned as shown; in addition, the peak at about − 5 p.p.m. is from internal ATP/ADP. (Adapted from Ogawa *et al.* 1978.)

to the mitochondria are significantly broader than those from the external medium, but nevertheless they are readily detectable. The inorganic phosphate also generates two signals because the internal and external pools of this compound are in environments of differing pH. These observations suggest that at 0 °C the detected metabolites exchange slowly (on the n.m.r. time scale) between internal and external media for otherwise a single exchange-averaged signal would be observed for each compound (see § 6.5). It may be possible to obtain approximate estimates for the rates of exchange over a range of temperatures using the theory outlined in § 6.5. For this and other reasons, it would be most interesting to perform similar studies at higher temperatures.

These observations provide no evidence of significant tight binding within the mitochondria of metabolites such as ADP, but equally they do not exclude this possibility. Such binding, if it exists, could at least partially explain some of the discrepancies between the concentration measurements made by n.m.r. and by freeze clamping (see § 3.1).

Aerobic rat liver cells also generate two inorganic phosphate signals which can be assigned to the cytoplasm and mitochondria (Cohen *et al*. 1978). Measurements of pH gradients in these cells are entirely consistent with Mitchell's chemiosmotic hypothesis of oxidative phosphorylation. Additional agreement with Mitchell's theory has been obtained from some ^{31}P n.m.r. experiments on *E. coli* (Ugurbil, Rottenberg, Glynn, and Shulman 1978). Following the addition of glucose to the cells, two inorganic phosphate signals corresponding to the pH environments internal and external to the cells are again observed. The magnitude of the pH gradient is reduced on addition of dicyclohexylcarbodiimide (DCCD), an inhibitor of the membrane-bound ATPase, or of the uncoupler *p*-trifluorometoxyphenyl-hydrazone (FCCP).

All of these experiments indicate that, at least in principle, the possibility exists of studying mitochondrial compartmentation in intact tissues and organs. In practice, however, because of the relative simplicity of cellular and sub-cellular systems and the ease with which they can be manipulated, it would seem logical for further work to be performed firstly on these simpler systems in order to characterize further the nature of the n.m.r. signals from the mitochondrial compartment. The versatility of these studies should be enhanced considerably by the use of sample chambers that enable cell preparations to remain viable for long periods of time at physiological temperatures (Balaban, Gadian, Radda, and Wong 1981*b*).

Storage systems, such as the catecholamine storage vesicle of the adrenal medulla, are particularly amenable to n.m.r. study. The vesicles of the adrenal medulla, known as chromaffin granules, contain high concentrations of catecholamine and ATP. Because of differences in composition and pH, ^{31}P n.m.r. signals from the intragranular ATP can be distinguished from the signals of external ATP (see Fig. 3.3). Interactions with the other compounds of the granules cause the titration curves for intragranular ATP to differ considerably from the standard curves shown in Fig. 2.10. Nevertheless, by using the appropriate calibrations it can be shown from the resonance frequencies of ATP that the internal pH in the resting granules is about 5.7 (Casey, Njus, Radda, and Sehr 1977; Njus *et al*. 1978; Pollard *et al*. 1979). This pH drops by about 0.5 pH units when ATP is added to isolated granules and is hydrolysed by the membrane-bound ATPase, and it has been possible to show that this ATPase is an inwardly directed electrogenic proton pump (Casey *et al*. 1977; Njus *et al*. 1978).

Similar experiments can be performed on blood platelets which store in membrane-bound granules biogenic amines such as 5-hydroxytryptamine, divalent metal ions such as Ca^{2+}, and nucleotides, mainly ATP and ADP. In many of their properties these granules are similar to chromaffin granules. However, the ^{31}P n.m.r. spectra of human platelets differ in that no high-resolution signals can be observed from the intragranular ATP (Ugurbil *et al*. 1979*a*; Costa *et al*. 1979). This is presumably because of the highly restricted mobility of ATP and ADP within the granules; possibly the divalent metal ions form aggregates of high molecular weight with ATP and ADP. It is interesting that the intragranular ATP and ADP are detectable in pig platelet granules, which indicates less immobilization than in human platelets. This presumably results from the known differences in composition between pig and human platelet granules (see Ugurbil *et al*. 1979*a*). In particular, the pig granules

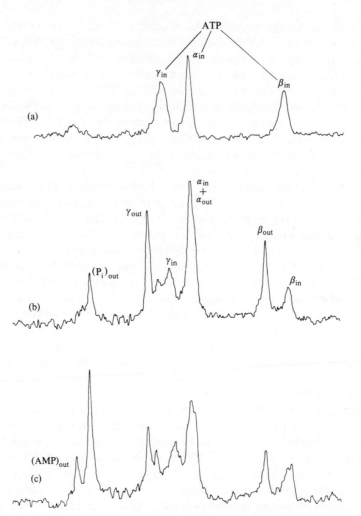

Fig. 3.3. ^{31}P n.m.r. spectra obtained at 129 MHz from a suspension of chromaffin granules at 25 °C. The suffices 'in' and 'out' refer to intragranular and extragranular ATP. (a) Prior to the addition of ATP; (b) 1–5 min after the addition of 100 mM ATP + 100 mM MgSO$_4$; (c) 34–8 min after the addition of ATP. After the addition of ATP, the γ_{in} signal shifts in a manner that indicates that the pH inside the granules has decreased by about 0.5 pH units. (From Casey *et al.* 1977. Adapted with permission from *Biochemistry* **16**, 972. Copyright © 1977 American Chemical Society.)

contain Mg^{2+} rather than the Ca^{2+} of human granules, and it is possible that this distinction between the two could explain the difference in aggregation properties.

3.3. Kinetic studies using ^{13}C n.m.r.

In many ways (see § 1.10) ^{13}C n.m.r. is better suited than ^{31}P n.m.r. for studying the detailed kinetics of metabolic pathways, and in this section we briefly discuss

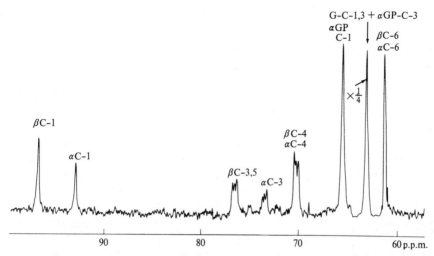

Fig. 3.4. ^{13}C n.m.r. spectrum obtained in 17 minutes at 90.5 MHz from a suspension of rat liver cells, 18–35 minutes after the addition of 22 mM (1,3-^{13}C) glycerol. The various glucose signals are assigned as shown to specific isomers and carbon positions. In addition, G-C-1,3 is the labelled glycerol peak, and αGP is α-glycerophosphate. (From Shulman *et al.* 1979. Copyright © 1979 by the American Association for the advancement of Science.)

some recent applications of ^{13}C n.m.r. which highlight the attractiveness of the technique.

Cohen, Ogawa, and Shulman (1979a) have used ^{13}C n.m.r. to follow the synthesis of glucose from glycerol in rat liver cells. Figure 3.4 shows a ^{13}C n.m.r. spectrum that they obtained from a suspension of rat liver cells after the addition of glycerol labelled with ^{13}C at the C-1 and C-3 positions. Signals can be observed, as expected, from glycerol, α-glycerophosphate, and glucose. The glucose signals can be readily assigned to specific carbons of the α- and β-anomers, and to a first approximation all of the ^{13}C label appears in the 1, 3, 4, and 6 carbons. Incorporation of the label into the C-4 and C-6 positions is expected on the basis of exchange between the triose phosphates, dihydroxyacetone phosphate and glyceraldehyde 3-phosphate (see Fig. 3.1). In fact, the equal labelling at all four positions implies that this reaction is at equilibrium or that other fluxes through the triose pool are negligible.

Spectra were obtained as a function of time in liver cells obtained from normal and hyperthyroid rats, and interesting differences were observed between the two types of cells. In particular, the rates of glycerol consumption and glucose synthesis were higher in the cells of the hyperthyroid rats, and moreover the concentration of α-glycerophosphate remained lower than in the normal cells. These observations can be interpreted in terms of the flux of α-glycerophosphate through the cytoplasmic and mitochondrial α-glycerophosphate dehydrogenases, and are totally consistent with measurements obtained *in vitro*, which had shown that the activity of mitochondrial α-glycerophosphate dehydrogenase is increased in cells from hyperthyroid rats.

Another interesting feature of the spectra is that the peaks from the C-3 and C-4 carbons of glucose appear as triplets. This is because a ^{13}C-labelled C-3 that is adjacent to an unlabelled C-4 will give rise to a single peak. However, if the labelled C-3 is adjacent to a labelled C-4, it will be split into a doublet as a result of the spin–spin coupling between the two ^{13}C nuclei. The net effect is that the C-3 resonance is split into three peaks corresponding to a central singlet and an outer doublet. Each spectrum therefore contains contributions from both labelled and unlabelled triose sources, and can thus be interpreted not only in terms of the amount of label incorporated into glucose but also in terms of the total glucose that is produced.

In an extension of this work (Cohen, Shulman, and McLaughlin 1979*b*) mouse livers have been perfused with ^{13}C-labelled alanine and ethanol, and the complexity of the spectra shown in Fig. 3.5 should be sufficient to illustrate the scope of the information that should be available from such studies. Signals can be observed from a wide range of metabolites, including glucose, aspartate, glutamate, glutamine, alanine, and lactate, and by observing the rate and extent of incorporation of label into the individual carbons of these molecules detailed information can be obtained about the relative rates of many of the reactions of gluconeogenesis.

For example, if we just consider the distribution of label that is observed in glucose, the 3- and 4-carbons are approximately equally labelled, suggesting that the

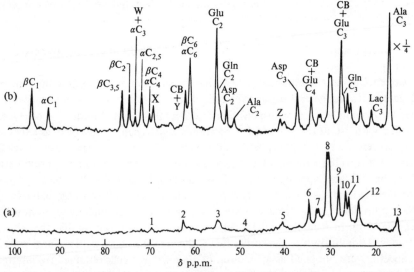

Fig. 3.5. ^{13}C n.m.r. spectra obtained at 90.5 MHz from a perfused mouse liver at 35°C. (a) ^{13}C natural abundance spectrum accumulated before the substrate was added. The peaks labelled 1, 2, 6, 7, 8, 9, 11, 12, and 13 have been assigned to the triglycerides of palmitic, oleic, and palmitoleic acids. 8 mM (3–^{13}C) alanine and 20 mM unlabelled ethanol were then added at 0 min and again at 120 min. (b) Spectrum obtained during the period 150–180 min. Peak assignments are as follows: βC_1, αC_1, $\beta C_{3,5}$, βC_2, αC_3, $\alpha C_{2,5}$, αC_4, βC_6, and αC_6 are the carbons of the two glucose anomers; Glu C_2, glutamate; Gln C_2, glutamine; Asp C_2, aspartate; Ala C_2, alanine; Lac C_3, lactate; CB, cell background peak; W, X, Y, and Z, unknowns. (Adapted from Cohen *et al.* 1979*b*.)

triose phosphate isomerase reaction is close to equilibrium. However, the enrichments at C-1 and C-2 are significantly lower than those at C-5 and C-6, as a result of activity of the pentose phosphate pathway. The C-5/C-2 and C-6/C-1 ratios are consistent with there being about 15 per cent flux through this pathway. There is no need to stress the power of this approach to the study of *in vivo* kinetics, and further publications consolidating and extending these studies are awaited with great interest. Of course, one of the main problems with ^{13}C n.m.r. is the expense of using very large amounts of suitably enriched materials. It is only to be hoped that a wider variety of reasonably cheap labelled compounds will become available in the future.

One intriguing development has been reported by Styles *et al.* (1979) who have modified a spectrometer in order to obtain ^{13}C and ^{31}P signals in alternate scans. In this way they have obtained simultaneous ^{13}C and ^{31}P time courses of metabolites in red blood cells in an attempt to establish the role of the 2,3-diphosphoglycerate bypass in these cells. This bypass, around the reaction catalysed by phosphoglycerate kinase, constitutes an important feature of red cell metabolism, but its role remains unclear (see Brown and Campbell 1980). Because of the complementary nature of the information provided by ^{13}C and ^{31}P n.m.r., there would be obvious attractions in performing similar studies on a range of other preparations.

3.4. ^{1}H n.m.r. studies

Because of the difficulties associated with high-resolution ^{1}H n.m.r. studies of living systems (see § 1.10), these studies are at an earlier stage in their development than corresponding investigations using ^{31}P or ^{13}C n.m.r. Nevertheless, there is considerable potential, as illustrated by the studies of red blood cells performed by Brown, Campbell, and co-workers (see Brown and Campbell (1980) for a review of this work). Signals can readily be identified from compounds such as haemoglobin, lactate, pyruvate, glucose, and glutathione (see Fig. 3.6). In addition to following the kinetics of, for example, lactate production, measurements can also be made of the internal pH, the oxidation state of the glutathione, the isotope exchange of lactate and pyruvate, and the transport of metabolites into the cells (see various sections of Chapter 2).

One of the most interesting features of this work is the study of $^{1}H/^{2}H$ exchange which can provide detailed information about the kinetics of specific enzyme systems within the red cells. For example, if protonated lactate and pyruvate are added to a suspension of the cells in $^{2}H_2O$ solvent, the ^{1}H n.m.r. signals from the methyl groups gradually decline (see Fig. 3.7) because of exchange with the solvent water. This process takes place via the following reactions:

$$\left[\begin{array}{c} CH_2^- \\ | \\ \text{haemoglobin}-N=C \\ | \\ CO_2^- \end{array} \right] \quad \leftarrow \quad \begin{array}{c} CH_3 \\ | \\ C=O \\ | \\ CO_2 \end{array} \underset{\substack{\text{lactate} \\ \text{dehydrogenase}}}{\rightleftharpoons} \begin{array}{c} CH_3 \\ | \\ HCOH \\ | \\ CO_2^- \end{array} .$$

The exchange of pyruvate methyl protons with solvent water is catalysed by protein $\alpha -NH_2$ groups, probably through the formation of transient Schiff's base

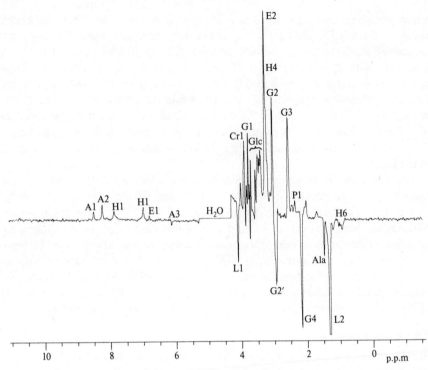

Fig. 3.6. ^{1}H n.m.r. spectrum of intact red blood cells obtained at 470 MHz. The spectrum shows signals from ATP (A), haemoglobin (H), ergothioneine (E), creatine (Cr), lactate (L), glucose (Glc), glutathione (G), pyruvate (P), and alanine (A). The cells were suspended in H_2O buffer, and the solvent peak was suppressed using the spin-echo radiofrequency pulse sequence, with a delay between the 180° and 90° pulses of 60 ms. The use of this pulse sequence removes most of the haemoglobin signals, and also causes some of the signals to be inverted (see § 6.3.7 for a discussion of the spin-echo sequence and its effects). (From the work of Brown and Campbell.)

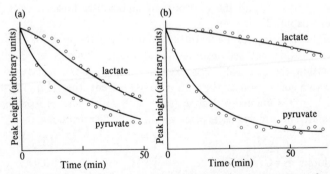

Fig. 3.7. Exchange of pyruvate and lactate methyl ^{1}H with solvent ^{2}H catalysed by human erythrocytes in ^2H_2O-containing Krebs–Ringer buffer. Peak heights of methyl resonances in ^{1}H spectra versus time (a) in normal cells, and (b) in cells following the addition of 4 mM oxalate. (From Brindle, Brown, Campbell, Foxall, and Simpson, 1980.)

links. From the data shown in Fig. 3.7 rate constants can be obtained for the lactate dehydrogenase reaction in untreated cells and also in cells for which this reaction is inhibited by the addition of oxalate (Brindle *et al.* 1980).

Exchange at the C-2 position of lactate can also be detected because it causes an inversion of the lactate methyl resonance in the spin-echo spectrum. The reason why some peaks are inverted in these spectra is given in § 6.3.7. Exchange at the C-2 position is of interest because it is dependent upon, and therefore provides information about, the activities of four glycolytic enzymes: aldolase, triose phosphate isomerase, glyceraldehyde 3-phosphate dehydrogenase, and lactate dehydrogenase.

So far, much of the emphasis in these studies of red cells has been to develop new methods of applying n.m.r. to living systems, and it will be interesting to see how the applications of these methods develop in the next few years. The non-invasive measurement of transport could be especially valuable, particularly if the method can be extended to investigate transport between different regions of the cell.

In addition to the difficulties associated with resolving the large number of resonances in ^{1}H n.m.r. spectra, there is also the problem that the water generates an enormous ^{1}H signal that can effectively 'swamp' the much smaller signals from metabolites. There are a number of ways of overcoming this problem, one of which involves a method of data collection known as correlation n.m.r. This technique, first described by Dadok and Sprecher (1974), has been successfully used by Ogino, Arata, and Fujiwara (1980) in their ^{1}H studies of the anaerobic metabolism of *E. coli* cells. They used (1-^{13}C) glucose as the sole carbon source, and were therefore able to observe ^{13}C–^{1}H spin–spin couplings in the ^{1}H spectra. By comparing the amounts of ^{13}C-labelled and unlabelled metabolites they could determine the relative fluxes through the glycolytic pathway and the pentose phosphate shunt, and it was concluded that as much as 22 per cent of the glucose was catabolized via the pentose shunt.

3.5. Activities of creatine kinase and phosphorylase in vivo

As discussed in § 2.5 and Appendix 6.1, it is now possible to use n.m.r. to measure enzyme activities *in vivo* under steady state conditions. One of the techniques that is employed is known as saturation transfer n.m.r. and its first *in vivo* application was in the measurement of ATPase activity in *E. coli* (Brown, Ugurbil, and Shulman 1977). Here we shall consider some more recent studies of creatine kinase activity in skeletal muscle and heart.

Creatine kinase catalyses the reaction

$$\text{phosphocreatine}^{2-} + \text{Mg ADP}^- + \text{H}^+ \rightleftharpoons \text{Mg ATP}^{2-} + \text{creatine}$$

which regenerates the ATP that is broken down during mechanical activity (see Fig. 3.8). Saturation transfer n.m.r. studies of anaerobic frog gastrocnemius muscles at 4 °C show that the forward (phosphocreatine to ATP) and reverse reaction fluxes are about 1.6 mM s^{-1} in resting muscle (Gadian *et al.* 1981). Since these rates are much faster than the rates of any other reactions utilizing ATP, the measurements

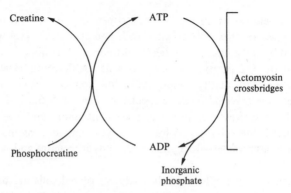

Net reaction: phosphocreatine → creatine + inorganic phosphate

Fig. 3.8. The main reactions that take place during muscular contraction.

confirm that as expected the reaction is close to equilibrium in resting muscle. This is an important observation, for it enables us to calculate the concentration of free ADP from the known equilibrium constant for this reaction together with the concentrations of the other reactants (all of which can be evaluated by combining n.m.r. and freeze-clamping data).

In contracting muscle it was found that there was a net breakdown of phosphocreatine of $0.75 \, \text{m s}^{-1}$. This corresponds to the difference between the forward flux (measured to stay at about $1.6 \, \text{mM s}^{-1}$) and the reverse flux (which declines to about $0.85 \, \text{mM s}^{-1}$). From these measurements it is clear that during contraction the creatine kinase reaction is no longer close to equilibrium, for contrary to expectation the forward and back rates differ by a factor of 2. Nevertheless, kinetic modelling shows that the measured rates are entirely consistent with the well-known observation that during contraction the ATP concentration falls by less than 2–3 per cent. Figure 3.9 shows results based on computer modelling that indicate how the metabolite levels change during a 3 s contraction. Note that, apart from the expected changes in phosphocreatine and inorganic phosphate, there is a marked increase in free ADP, which rises from about $20 \, \mu\text{M}$ to $60 \, \mu\text{M}$ during the 3 s contraction. It is this threefold increase in ADP that drives the reaction forward during contraction, and we can now see how it is that, although the reaction is not close to equilibrium, it is nevertheless able to maintain the ATP level almost constant. The reason is that it only needs a 1 per cent drop in ATP (from $4 \, \text{mM}$ to $3.96 \, \text{mM}$) to produce a threefold increase in ADP (from $20 \, \mu\text{M}$ to $60 \, \mu\text{M}$). Such a large percentage increase in ADP tends to drive the creatine kinase reaction from left to right to maintain the ATP level almost constant.

The situation for perfused beating hearts is more complex than in resting muscle, partly because of the relatively high rate of ATP utilization; there is a range of reactions utilizing and synthesizing ATP that could have activities comparable with that of creatine kinase, and this greatly complicates interpretation of the measurements. In the n.m.r. experiments the heart remains in a steady state, and yet the

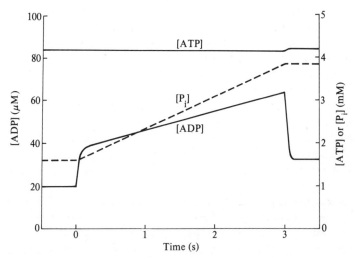

Fig. 3.9. Computer modelling of the changes in concentration of free ATP, ADP, and inorganic phosphate resulting from a 3-s contraction. The muscles were contracting during the time interval 0–3 s. The decline in the phosphocreatine level is not shown, but it matches almost exactly the rise in inorganic phosphate. (Adapted from Gadian *et al.* 1981.)

flux measured by n.m.r. for the forward reaction of creatine kinase seems somewhat larger than the reverse flux (Nunnally and Hollis 1979; Garlick 1979; Ackerman, Bore, Gadian, Grove, and Radda 1980*a*). This apparent anomaly could be interpreted in terms of the reactions that compete for ATP (e.g. ATP hydrolysis and synthesis or adenylate kinase activity), or alternatively in terms of compartmentation of the metabolites involved in the creatine kinase reaction (Nunnally and Hollis 1979). Further experiments are required to establish whether these explanations are correct. However, it is interesting to note that there is also a small discrepancy between the forward and back fluxes measured in the resting frog skeletal muscle preparation. (These were quoted above as being about $1.6 \, \text{mM s}^{-1}$, but the measured forward flux was somewhat larger than the measured reverse flux.) This discrepancy could perhaps be explained in terms of adenylate kinase activity within the muscle, and perhaps this reaction is also partially responsible for the discrepancy in the heart measurements. These uncertainties can be regarded as being drawbacks of the method, but equally they highlight the fact that additional information can in principle be obtained.

Phosphorylase is another important enzyme whose *in vivo* activity can be investigated by n.m.r. Phosphorylase catalyses the phosphorolytic cleavage of the 1,4-glycosydic bonds of glycogen to form glucose 1-phosphate:

$$\text{glycogen}_{(n\text{-glucose})} + \text{P}_i \rightarrow \text{glucose 1-phosphate} + \text{glycogen}_{(n-1)}$$

The glucose 1-phosphate is converted via the enzyme phosphoglucomutase to glucose 6-phosphate, and thereby enters the main pathway of glycolysis.

Phosphorylase activity is subject to extensive control. The enzyme exists in two forms, *a* and *b*, which can be interconverted by specific enzymes (see for example Newsholme and Start 1973), and there are three main mechanisms whereby the activity of the enzyme is controlled. Firstly, phosphorylase *b*, which is the form that is present in resting muscle, is controlled by metabolite levels in such a way that utilization of metabolic energy by mechanical activity stimulates phosphorylase *b* activity. The other two mechanisms both involve conversion of the *b* form to the active *a* form. The conversion of phosphorylase *b* to phosphorylase *a* is stimulated by adrenaline and by the release of Ca^{2+} ions that also activate muscular contraction. Numerous studies have been performed of phosphorylase *in vivo* and *in vitro*, and n.m.r. can provide some complementary information about the activity of the enzyme *in vivo*.

For example, hearts perfused with 2-deoxyglucose accumulate 2-deoxyglucose 6-phosphate, for this compound is formed within the heart but can only be metabolized very slowly. 2-deoxyglucose 6-phosphate inhibits phosphorylase *b*, and in recent n.m.r. studies it has been possible to compare the rates of glycogen breakdown in the presence and absence of this compound (Bailey *et al.* 1981). It was concluded that in the rat heart, after 5 min of global ischaemia, the *b* form of phosphorylase contributes a significant amount (about 50 per cent) of the total activity of this enzyme.

In contrast, studies of frog gastrocnemius muscles at 4 °C suggest that the predominant mechanism of phosphorylase activation is via Ca^{2+}-controlled conversion of phosphorylase *b* to phosphorylase *a* (Dawson, Gadian, and Wilkie 1980*b*). A deeper insight into phosphorylase activity *in vivo* should be obtained by using n.m.r. to compare the metabolism of normal and I-strain mice. I-strain mice carry a genetic deficiency that makes it impossible for them to convert phosphorylase *b* to phosphorylase *a*. This strain of mice therefore relies entirely on phosphorylase *b* activity for its breakdown of glycogen. N.m.r. studies of these mice should provide a perfect example of the way in which n.m.r. can complement the more traditional approaches to metabolic studies.

Finally, the low levels of inorganic phosphate and free AMP could explain why the activity of phosphorylase *b* in resting muscle is considerably less than the value predicted from studies on the isolated enzyme combined with metabolic measurements by freeze extraction (see Busby and Radda 1976).

3.6. Relationships between biochemical state and physiological function

A particularly attractive feature of n.m.r. is that metabolism can be monitored continuously in a physiologically functioning tissue or organ. For this reason, n.m.r. provides an ideal method of relating metabolic state to physiological function, as exemplified by some ^{31}P n.m.r. studies of muscular fatigue, a subject about which surprisingly little is known.

Following earlier studies of frog sartorius muscles (1977*a*), Dawson *et al.* (1978, 1980*a,b*) have studied the biochemical basis of fatigue in frog gastrocnemius muscles maintained at 4 °C under anaerobic conditions. Figure 3.10 shows the results of a typical experiment in which frog gastrocnemii were stimulated re-

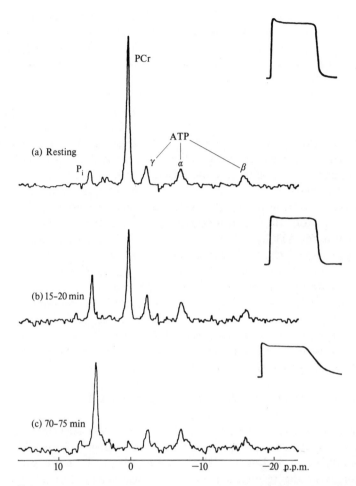

Fig. 3.10. ^{31}P n.m.r. spectra obtained at 129 MHz from anaerobic frog gastrocnemii at 4 °C during a fatiguing series of 5 s contractions repeated every 5 min. Adjacent to the spectra are mechanical records showing the time course of isometric force development; the force progressively declines and relaxation becomes slower. (Adapted from Dawson *et al.* 1980*a*.)

petitively for 5 s every 5 min. The force measurements shown at the right of each spectrum reveal the typical signs of fatigue; the force gradually declines in magnitude and the rate of mechanical relaxation becomes progressively slower. The spectra indicate the expected decline in phosphocreatine, increase in inorganic phosphate, and decline in pH due to the formation of lactic acid.

An interesting conclusion from this work was that, throughout the course of fatigue, the force developed by the muscles remained proportional to the rate at which ATP was hydrolysed during contraction. This suggests that any factors that reduce the rate of hydrolysis of ATP would reduce the force correspondingly and hence result in fatigue. It now remains to establish the precise nature of these factors. One possibility is that the build up of ADP, inorganic phosphate, and H^+ (from lactic acid formation) could, by product inhibition, slow down the rate of ATP hydrolysis at the muscle cross-bridges. For obvious reasons, lactic acid accumulation has in the past been considered to be a cause of fatigue. However, as these n.m.r. studies demonstrate, lactic acid accumulation will generally be accompanied by the net conversion of high-energy phosphates to low-energy phosphates, and it is bound to be difficult to distinguish between the effects that these two processes have on fatigue. It is interesting to note that patients with McArdle's syndrome generate very little lactic acid (see § 4.1.3), but fatigue rapidly during exercise; this suggests that the changing phosphorylation state of the muscles, rather than lactic acid accumulation, may be the important factor in the onset of fatigue.

Another manifestation of fatigue is the decline in the rate of mechanical relaxation following electrical stimulation, which can clearly be seen in the force records of Fig. 3.10. From the n.m.r. measurements it is possible to calculate the free energy change for ATP hydrolysis *in vivo*, and it was suggested that the decline in free energy of ATP hydrolysis that is observed to accompany fatigue causes a reduction in the rate of Ca^{2+} uptake into the sarcoplasmic reticulum, which in turn causes a slowing down in the rate of mechanical relaxation.

It is not a trivial matter to oxygenate skeletal muscle preparations that are suitable for n.m.r. experiments, and these studies of muscular fatigue, for which no oxygen is desired, provide an ideal way of circumventing the problem. Recently, Brown and Kushmerick (1981) have succeeded in performing some n.m.r. studies of perfused cat biceps muscle, which they were able to maintain in good metabolic condition for periods of many hours. An alternative approach is to allow the animal itself to do the perfusion, i.e. to study the muscle within the live animal, using the methods described in Chapter 4.

Perfusion of the heart, kidney, and liver is rather more straightforward, although one always has to beware of deficiencies in oxygen supply. Many of the methods have been described by Ross (1972) and can be adapted for n.m.r. studies. Several investigations of the relationship between cardiac function (e.g. left ventricular pressure) and metabolic state have been reported, including studies on acidosis (Jacobus *et al.* 1978), heart rate (Ackerman *et al.* 1980*a*), and the effects of several different inotropic agents (Dube *et al.* 1981). In the acidosis studies the intracellular pH within the heart was reduced in two different ways: either by reducing the pH of the perfusate, or by temporarily stopping the flow of perfusate into the heart (a condition known as ischaemia). The aim of the experiments was to evaluate the importance of pH in determining cardiac function. It was shown that, for a given decline in pH, cardiac function was impaired more by ischaemia than by simply reducing the external pH. During ischaemia the concentrations of high-energy phosphates rapidly decrease, and the results suggest that the decline in function resulting from ischaemia is only partially due to the decline in pH; there must be other metabolic effects — perhaps the increase in inorganic phosphate or ADP — that contribute to the decline in cardiac function. In fact, a variety of studies on skeletal muscle, heart, and kidney suggest that pH is just one among a number of factors that govern tissue function and viability.

3.7. Preservation and viability studies

The acidosis experiments described above relate closely to investigations of how tissues can be protected from excessive damage during prolonged periods of ischaemia. The main purpose of such investigations is to develop better methods of preserving organs that are subjected to periods of ischaemia during surgery; one particular application relates to organs to be used in transplantation surgery. It is still unclear what biochemical factors lead to the irreversible damage that results from prolonged periods of ischaemia; possibilities include low pH and depletion of the adenosine phosphate pool. N.m.r. is ideally suited to investigating this important problem, and some interesting results are beginning to emerge.

The extent of recovery from ischaemia is affected by temperature, the duration of the ischaemia period, and the use of agents such as KCl that arrest the heart. Hollis (1979) has described some studies in which a dose of KCl cardioplegic (heart-arresting) solution was administered to hearts at the time of initiating ischaemia. The main effects of KCl administration were to reduce the decline in pH and ATP concentration during the ischaemic period, and during recovery to enhance the recovery of ATP and the performance of the heart in comparison with the control (no KCl) experiments.

Garlick, Radda, and Seeley (1979) have studied the protective effects of perfusing hearts with a medium of high buffering capacity prior to ischaemia. They found that pre-treatment of hearts with a buffer containing 100 mM Hepes (4-(2-hydroxyethyl)-1-piperazine-ethanesulphonic acid) does indeed provide significant protection. During subsequent ischaemia, the intracellular pH and the ATP and phosphocreatine concentrations declined more slowly than in the control, and complete metabolic recovery was observed on re-perfusion after 30 min of ischaemia at 37 °C. In contrast, there was no metabolic recovery in the control.

Studies have also been performed to evaluate the importance of pH in renal preservation (Bore *et al.* 1981). These studies are of direct clinical significance because a flushing solution is commonly used during donor nephrectomy prior to kidney transplantation. Flushing solutions of widely differing composition are currently in clinical use, and the aim of this n.m.r. work was to evaluate the importance of the buffering capacity of the flushing solutions in preserving kidney function during warm ischaemia. It was found that the buffering capacity of the flushing solution is indeed an important parameter to consider when designing such solutions; pre-treatment with a buffered flushing solution produced significant improvements in animal survival studies.

4

Applications to animals and human beings

Over the last eight years, there has been increasing interest in an n.m.r. technique known as 'spin-imaging', or 'zeugmatography'. This technique involves the use of magnetic field gradients to provide information about the spatial distribution of molecules within a sample. Various ingenious methods of spin-imaging have been devised, but in applying these methods there has been a common underlying theme; interest has centred on obtaining 2- or 3-dimensional images of the proton signals that are generated by human beings. The images generally reflect the distribution of mobile protons contained within water and fats, and can provide remarkably clear discrimination between different tissues, as illustrated by the head scan that was shown in Plate 1. It is now apparent that ^{1}H spin-imaging could complement existing methods of medical imaging such as ultrasound and X-ray CT scanning, and later in this chapter we discuss some of the possible clinical applications.

For the biochemist and the clinician, it would undoubtedly be of great interest to combine a spin-imaging method with high-resolution n.m.r., in order to evaluate the metabolic state of different tissues within an animal or human being. However, it should be noted that the metabolites observed in high-resolution n.m.r. studies have typical concentrations of about 1–5 mM, whereas the protons observed in spin-imaging are present at concentrations of up to about 100 M. Furthermore, ^{13}C and ^{31}P n.m.r. are inherently less sensitive than ^{1}H n.m.r., and for these reasons there would be severe sensitivity problems associated with spin-imaging of ^{13}C- or ^{31}P-containing metabolites. Acceptable signal-to-noise ratios could only be obtained at the expense of a dramatic reduction in spatial resolution; thus one would be limited to relatively crude images, consisting perhaps of at most 20 volume elements within the region of interest.

A second problem associated with ^{13}C or ^{31}P spin-imaging is that the preservation of high-resolution information is not compatible with many of the imaging methods that are in use. This particular problem has been overcome by Bendel, Lai, and Lauterbur (1980), who have managed to obtain ^{31}P n.m.r. images of phantom objects containing inorganic phosphate, ATP, and phosphocreatine within different compartments. However, they stress the difficulties associated with the inherent insensitivity of the method, and conclude that because of this, ^{31}P spin-imaging is not practical for imaging of human beings, at least at the present stage of technological development.

Although spin-imaging of metabolites does not appear to be feasible, it is not impractical to investigate the metabolic state of a selected, localized region within an animal or human being, and two methods of localization have been successfully used for ^{31}P studies of animals and human limbs; at the time of writing, however, ^{13}C studies of a similar nature have only just begun. In the first half of this chapter, we describe these methods of localization and the results that have been obtained; the latter half of the chapter deals with spin-imaging and its applications.

4.1 Metabolic studies of animals and humans

4.1.1 *The use of surface coils*

One method of localization make use of an unusual type of radiofrequency coil, which has been termed a surface coil (Ackerman *et al.* 1980*b*). The special properties of such coils are described in detail in § 8.4.2. Here, it is sufficient to point out that if a surface coil is placed adjacent to any object, it will under normal circumstances detect signal from an approximately disc-shaped region of the object immediately in front of the coil, of radius and thickness approximately equal to the radius of the coil. A surface coil therefore provides a suitable, and remarkably simple method of localizing on a region that is close to the surface of an object. The experimental procedure is relatively conventional, the only modification being that the normal solenoidal or saddle-shaped radiofrequency coil is replaced by a surface

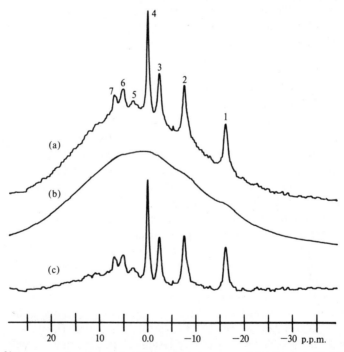

Fig. 4.1. ^{31}P n.m.r. spectra of the brain of an intact rat obtained at 73.8 MHz using a surface coil. (a) Original brain spectrum accumulated over 60 min. (b) Same as (a) except that the narrow components have been removed by applying an exponential multiplication corresponding to a line broadening of 400 Hz (See § 7.4.2). (c) The difference spectrum (a) minus (b). Peak assignments are as follows: 1, 2, and 3, the β- , α- , and γ- phosphates of ATP; 4, phosphocreatine; 5, phosphodiesters, 6. inorganic phosphate and the 2-phosphate of 2, 3-diphosphoglycerate; 7, sugar phosphates (+ AMP, IMP), and the 3-phosphate of 2, 3-diphosphoglycerate. (From Ackerman *et al.* 1980*b* Reprinted by permission from *Nature* **283**, 167−70. Copyright © 1980 Macmillan Journals Limited.)

coil. Surface coils are ideally suited for examining the metabolic state of skeletal muscle *in vivo*, as illustrated by the studies of anaesthetized rats that were described in § 1.9.2, and by the subsequent studies of the human forearm to be discussed in § 4.1.3.

Surface coils are also of value for metabolic studies of the brain. Figure 4.1 (a) shows a spectrum obtained when a surface coil is placed against the head of an anaesthetized rat. The spectrum contains narrow signals from the metabolites within the brain, superimposed on the much broader signal from the immobile phosphate in the bone. It is a simple matter to remove the broad component (using the technique of convolution difference described in § 7.4.2), leaving just the narrow signals of the brain, as shown in Fig. 4.1 (c). Metabolic studies of the brain are notoriously difficult, because of the problems associated with perfusion and freeze-extraction, and therefore spectra such as that shown in Fig. 4.1 (c) can provide a large amount of new and useful information. It is immediately apparent that the level of inorganic phosphate is much less than that of ATP, in contrast with measurements quoted in the literature. In fact, precise quantitation of the peaks at about 5.0 and 6.5 p.p.m. is made difficult by the possible presence of two signals from the 2, 3-diphosphoglycerate that is present in the blood. This compound produces signals at these two positions, and so the peak at 5 p.p.m. could contain contributions from both inorganic phosphate and the 2-phosphate of 2, 3-diphosphoglycerate. However, even if the dominant contribution to this peak were from inorganic phosphate, the concentration of this compound deduced from the spectra would still be much less than anticipated from the literature values. Moreover, the ratio of signal intensities at −2.5 and −16 p.p.m. is very close to unity, indicating that ADP is not detectable (see § 2.1). Again, this is contrary to expectations based on literature values. These considerations lead to the conclusion that the phosphorylation potential in brain tissue is at least an order of magnitude higher than the generally accepted value of about 3400 M^{-1}.

In an extension of this work, the localizing properties of surface coils have been utilized to study the metabolic effects of regional ischaemia in the gerbil brain (Thulborn, du Boulay, and Radda 1981*a*). Following occlusion of the right carotid artery, spectra were monitored from different regions of the brain. As expected, the metabolic state deteriorated far more extensively in the right anterior hemisphere than in the left hemisphere and in the cerebellum. It is interesting to note that the concentrations of phosphocreatine and ATP appeared to decline in unison in the ischaemic region, in contrast to the situation in skeletal muscle, where there is very little change in the ATP level until the phosphocreatine is almost completely depleted. A similar observation regarding the phosphocreatine and ATP levels was made by Norwood *et al.* (1979) in their studies of the perfused neonatal rat brain, and this clearly requires further investigation.

Surface coils can be used to study internal organs such as the kidney, if coupled with surgery (Bore *et al.* 1981). However, there are obvious disadvantages of such an approach, and there is a clear need for developing a technique that could localise on internal organs without there being any requirement for surgery. One technique that has been successfully used has been termed topical magnetic resonance (t.m.r.).

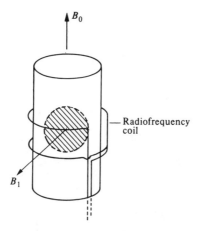

Fig. 4.2. Schematic diagram illustrating the effect of a t.m.r. experiment. The B_0 field is homogeneous over an approximately spherical volume, indicated by the central shaded region, but is very inhomogeneous outside this region. The radiofrequency coil therefore detects high resolution signals only from the shaded region. The t.m.r. technique was developed by Oxford Research Systems.

4.1.2 *Topical magnetic resonance (t.m.r.)*

T.m.r. employs a special type of field homogeneity coils, which profile the B_0 field in such a way that the field is very homogeneous over a central, approximately spherical volume, but elsewhere is very inhomogeneous (Fig. 4.2). As a result, high resolution signals are observed only from the central region.

The liver is very suitable for observation by t.m.r. Its lack of phosphocreatine provides a convenient marker for ^{31}P n.m.r. studies. Also, the unusually short T_1 values of its ^{31}P resonances (McLaughlin, Takeda, and Chance 1979; Iles, Griffiths, Sevens, Gadian, and Porteous 1980) provide an additional way of distinguishing liver signals from signals of other tissues. Figure 4.3 shows the results from an experiment designed to localize on the liver of a live, anaesthetized rat (Gordon, *et al*. 1980). The aim of the experiment was to obtain a spectrum from the live rat similar to that of a perfused liver (Fig. 4.3 (e)). The top spectrum was obtained from the rat in the absence of field profiling. It represents a summation of signals from skeletal muscle, liver, diaphragm, etc., and bears little resemblance to Fig. 4.3 (e). On applying the field profiling (Fig. 4.3 (b)), the resemblance becomes far better, as would be expected. The experiment was then repeated, but this time the radiofrequency pulses were applied at intervals of 220 ms, rather than the more customary 2 seconds. This rapid pulsing causes the signals from tissues other than the liver to be selectively saturated, as can be seen by comparing spectra (a) and (c). Fig. 4.3 (d) was obtained with rapid pulsing in the presence of field profiling, and now the spectrum bears a close resemblance to Fig. 4.3 (e). These observations, together with further studies that were performed, confirm the feasibility of

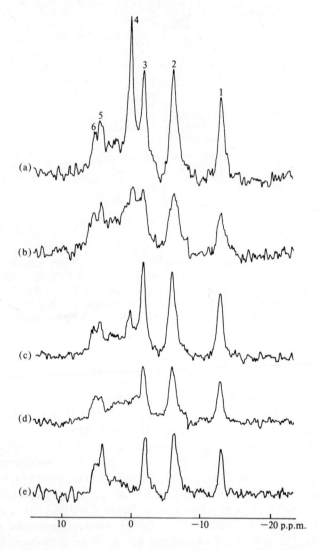

Fig. 4.3. ^{31}P n.m.r. spectra obtained at 73.8 MHz. Spectra (a)–(d) were obtained from an intact rat, and (e) was from a perfused rat liver. Peak assignments are as follows: 1, 2, and 3, the β- , α- , and γ-phosphates of ATP; 4, phosphocreatine; 5, inorganic phosphate; 6, sugar phosphates, AMP, IMP. Peaks 5 and 6 may also contain contributions from 2, 3-diphosphoglycerate in the blood. (a) Spectrum obtained using 128 90° pulses applied at intervals of 2 s in the absence of localizing fields. (b) Spectrum obtained using 128 90° pulses at intervals of 2 s with a homogeneous radius of 1 cm. (c) Spectrum obtained using 1024 90° pulses at intervals of 220 ms in the absence of localizing fields. (d) Spectrum obtained from 1024 90° pulses at intervals of 220 ms with a homogeneous radius of 1 cm. (e) Spectrum of a perfused liver obtained using 480 90° pulses at intervals of 220 ms. The convolution difference technique was used for all spectra, with line broadenings of 16 Hz and 233 Hz (see § 7.4.2). (From Gordon *et al.* 1980. Reprinted by permission from *Nature* **287**, 367–8. Copyright © 1980 Macmillan Journals Limited.)

localizing on the liver. In an extension of this work, it has been shown that, by combining the use of surface coils and t.m.r. it is possible to localize on the kidney of a live, intact rat (Balaban, Gadian, and Radda 1981*a*; see also § 8.4.2.).

T.m.r. is admittedly inefficient in the sense that information is thrown away from the outer regions, and certainly it would be preferable to image the whole sample rather than to focus on just one region. However, even if a suitable ^{31}P imaging technique were developed (with the associated sensitivity problems outlined earlier in the chapter), it would still rely on the B_0 field being homogeneous over the whole of the region of interest. This in itself would constitute a major problem in magnet design when dealing with very large samples. Therefore, at the present stage of technological development, we have to settle for the simple approach of localizing by means of surface coils or t.m.r., or a combination of the two.

4.1.3 N.m.r. studies of the human forearm

The n.m.r. methods that have been described in the previous sections can readily be used for studies of human metabolism, the main requirement being that a high resolution magnet must be available that is large enough to accommodate a human being, or at least an arm or leg. Magnets with a working bore of 20 cm are now in use for this type of work, and in this section we describe some of the recent studies

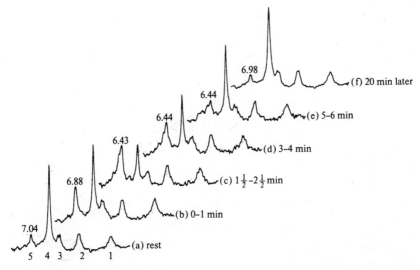

Fig. 4.4. ^{31}P n.m.r. spectra obtained at 32.5 MHz from a human forearm. The first spectrum (a) was recorded at rest, prior to exercise; subsequent spectra (b)–(f) were recorded during the periods shown, where 0 min corresponds to the time at which ischaemic exercise was started. Exercise was maintained during the period 0–1½ min, but arterial occlusion was maintained up to 3 min. Arterial flow was restored after this period. The signals are assigned as follows: 1, 2, and 3, the β-, α-, and γ-phosphates of ATP; 4, phosphocreatine; 5, inorganic phosphate. Measured pH values are given above each inorganic phosphate signal. (From Ross *et al.* 1981. Reprinted by permission of *New Eng. J. Med.* **304**, 1338 (1981).)

that have been carried out on human subjects. As we have seen, [31]P n.m.r. is ideally suited to studying muscle metabolism and the response to exercise, and we shall describe here how some [31]P n.m.r. studies of human exercise have formed part of a clinical diagnosis (Ross *et al.* 1981).

Figure 4.4 (a) shows a [31]P n.m.r. spectrum obtained in about 2 min from tissue within the flexor compartment of the forearm of a healthy subject. The spectrum contains signals as anticipated from ATP, phosphocreatine, and inorganic phosphate within the muscle. The inorganic phosphate level is rather higher than might be predicted from animal studies (see § 3.1), but nevertheless is much lower than the levels obtained by the technique of needle biopsy, presumably because this latter method involves an unavoidable breakdown of high-energy phosphates prior to analysis (see Cresshull *et al.* 1981). The intracellular pH deduced from the chemical shift of the inorganic phosphate signal is 7.04 (see § 2.2 for a discussion of the accuracy of pH measurements made by [31]P n.m.r).

Little change in the spectrum is observed during 5 min of complete arterial occlusion, produced by a sphygmomanometer cuff placed around the upper arm. On ischaemic exercise, however, dramatic changes are observed, as shown in Fig. 4.4 (b) and (c); the phosphocreatine declines, the inorganic phosphate increases, and the pH falls to around 6.5. In addition, the inorganic phosphate signal broadens, probably because there is a distribution of pH within the muscle tissue that is being examined. Following the restoration of arterial flow, the metabolic state recovers to

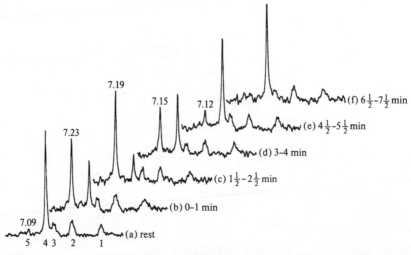

Fig. 4.5. [31]P n.m.r. spectra obtained at 32.5 MHz from a patient with McArdle's syndrome. The first spectrum (a) was recorded at rest, prior to exercise; subsequent spectra (b)–(f) were recorded during the periods shown, where 0 min corresponds to the time at which ischaemic exercise was started. Exercise was maintained during the period 0–¾ min, but arterial occlusion was maintained up to 3 min. Arterial flow was restored after this period. The signal assignments are as for Fig. 4.4. Measured pH values are given above each inorganic phosphate signal. (From Ross *et al.* 1981 Reprinted by permission of *New Eng. J. Med.* **304**, 1338 (1981).)

normal in a few minutes. Clearly, a considerable amount of information about muscle metabolism can be obtained remarkably rapidly; a complete study of this type takes only about 30 min, and the results are available almost immediately. Furthermore, the time resolution of the measurements can certainly be improved, particularly if the collection of data is synchronised with repeated muscular contractions.

It should be stressed that these measurements are totally non-invasive; the subject simply has to sit down, place his arm within the bore of a horizontal magnet, and perform any arm exercise that is asked of him. The question of safety is discussed in § 4.4; here, it is sufficient to say that there is no known hazard associated with the measurements.

Figure 4.5 shows spectra recorded in a similar manner from a patient who was suspected to have McArdle's syndrome. This is an inborn error of metabolism caused by lack of phosphorylase activity in skeletal muscle, and is diagnosed by the demonstration that ischaemic exercise fails to generate lactic acid. (The role of phosphorylase is illustrated by Fig. 3.1.) The main feature that distinguishes the spectra of Fig. 4.5 from those of the healthy subject shown in Fig. 4.4 is that there is no decrease in intracellular pH associated with ischaemic exercise; indeed, there is a small increase in pH. This observation is entirely consistent with the absence of activation of phosphorylase, and confirmed the diagnosis of McArdle's disease; the absence of phosphorylase was then firmly established by direct enzymatic assay of a muscle biopsy.

Studies of human limbs can also be performed using ^{1}H and ^{13}C n.m.r. It is interesting to note that ^{1}H n.m.r. could provide a method of estimating the relative amounts of muscle and fatty tissue, for water and fats produce ^{1}H signals that differ in chemical shift by about 2.5 p.p.m. These signals can readily be resolved with a high resolution spectrometer of the type that is used for ^{31}P and ^{13}C studies. In contrast, the ^{1}H imaging studies that are described below discriminate between different tissues, not on the basis of chemical shifts, but on the basis of their proton concentrations and relaxation times.

Fatty tissue can also be detected by ^{13}C n.m.r., without any requirement for isotopic enrichment. Preliminary studies of the human forearm have demonstrated that well-resolved ^{13}C signals can be observed from different types of carbon atoms within triglycerides and mobile membrane lipids (Alger *et al.* 1981).

4.1.4 Clinical applications of ^{31}P n.m.r.

The examination described above was probably the first ^{31}P n.m.r. study that has been made on a patient, and was almost certainly the first to form part of a clinical diagnosis. Therefore, it is far too early to predict how useful ^{31}P n.m.r. will be to the clinician. However, several types of application can be visualized. For example, n.m.r. can be used to assess the metabolic state of an organ such as a kidney, prior to transplantation. In this way, it may be possible to correlate the metabolic state of an organ with its subsequent function, and hence use n.m.r. as a method of predicting function. The preliminary results look most encouraging (Chan *et al.* 1981).

^{31}P n.m.r. could well be responsive to the effects of vascular disorders (e.g. heart attack and stroke). If so, the technique could be used to assess the size of the affected regions and the extent of metabolic damage, and also to monitor their response to medical treatment. Here, one would be taking full advantage of the harmless and non-invasive nature of the technique, for there should be no problems associated with making repeated observations over a prolonged period of time. There are also many metabolic disorders (e.g. specific enzyme deficiencies of the type described above) into which n.m.r. could clearly provide a most invaluable insight. It seems likely that the spectra of dystrophic muscle will differ from those of healthy muscle, and it may therefore be possible to use ^{31}P n.m.r. as a non-invasive method of monitoring how dystrophic muscle responds to the administration of drugs. Finally, it could be particularly useful to combine metabolic measurements using ^{31}P n.m.r. with measurements of blood flow and of other parameters that may be accessible to ^{1}H n.m.r.

The early ^{31}P n.m.r. results seem very promising, but it will be a few years before the true value of n.m.r. to the clinician becomes apparent. It should, of course, be remembered that apart from the direct value of n.m.r. in the clinic, the study of animal models using n.m.r. should enhance our understanding of human metabolism and of the metabolic basis and effects of clinical disorders.

4.2 Spin-imaging

4.2.1 *The methods of spin-imaging*

For most n.m.r. experiments, including all of those described in Chapter 3 and the first half of this chapter, one aims to optimise the homogeneity of the field B_0, for this ensures that the signals are as narrow as possible. In contrast, spin-imaging experiments require that the field be inhomogeneous, for different parts of the sample can then be characterised by the particular magnetic field they experience, and hence by the frequency of the signals that they generate.

As an example (see Fig. 4.6), consider a rectangular breaker of water which, if placed in a uniform magnetic field B_0, would produce a narrow ^{1}H signal. Let us now suppose that a small linear field gradient is superimposed upon B_0, so that the field is greater at the left side of the breaker than at the right. Since the resonance frequency of the protons is directly proportional to the field that they experience, the water at the left will generate a signal of higher frequency than the water at the right. An intermediate frequency will be detected from water in the centre. The resulting signal will therefore consist of a superposition of a continuous range of slightly shifted resonances, the resulting effect being that a rectangular-shaped signal is observed (Fig. 4.6 (a)). If the beaker had a circular cross-section, we would obtain the signal shown in Fig. 4.6 (b).

Suppose now that a plastic strip containing no protons is placed across the beaker, as shown in Fig. 4.6 (c). The resulting signal indicates that there are no protons within the region of the strip. Clearly, all of these n.m.r. signals provide information about the spatial distribution of water within the sample. However, a one-dimensional profile or projection that is obtained in this way provides a far from complete description of the structure of the sample, for similar signals could be obtained from samples of a variety of shapes.

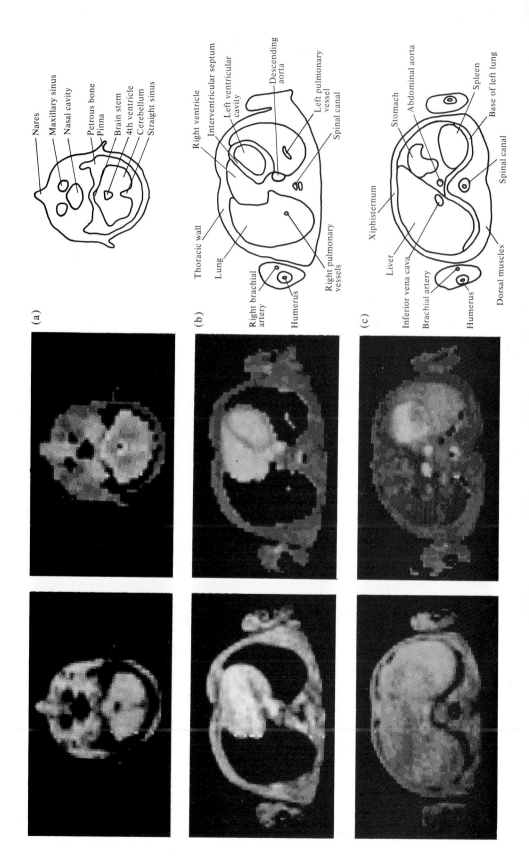

(a)

Nares
Maxillary sinus
Nasal cavity
Petrous bone
Pinna
Brain stem
4th ventricle
Cerebellum
Straight sinus

(b)

Right ventricle
Interventricular septum
Left ventricular cavity
Descending aorta
Left pulmonary vessel
Spinal canal
Thoracic wall
Lung
Right brachial artery
Humerus
Right pulmonary vessels

(c)

Stomach
Abdominal aorta
Spleen
Base of left lung
Spinal canal
Xiphisternum
Liver
Inferior vena cava
Brachial artery
Humerus
Dorsal muscles

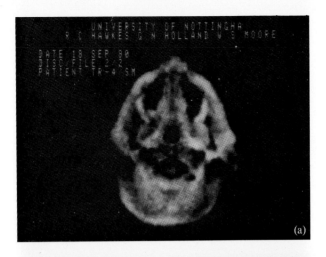

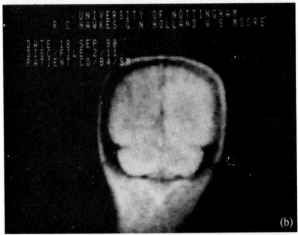

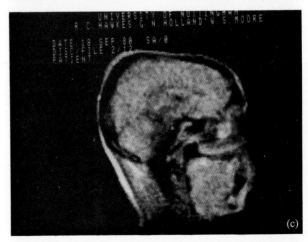

Plate 2. Head scans obtained by ^{1}H n.m.r. spin imaging. (a) Axial transverse, (b) coronal, and (c) sagittal views. (From the work of Hawkes, Holland, and Moore, Nottingham University.)

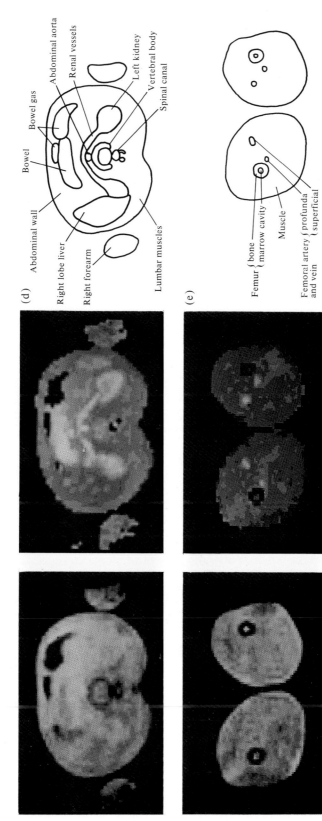

(d) Abdominal wall

Bowel gas
Bowel

Abdominal aorta
Renal vessels

Left kidney
Vertebral body
Spinal canal

Right lobe liver
Right forearm

Lumbar muscles

(e) Femur { bone
            marrow cavity

Muscle

Femoral artery { profunda
                 superficial

and vein

Plate 3. Body scans obtained by ¹H n.m.r. spin imaging. The three columns are respectively proton density images, spin-lattice relaxation time (T_1) images, and outlines of the important features. (a) Head section 25 mm below eyes; (b) chest section through heart; (c) abdominal section; (d) abdominal section; (e) thigh section. (From Edelstein, Hutchinson, Johnson, and Redpath 1980, Copyright of the Institute of Physics.)

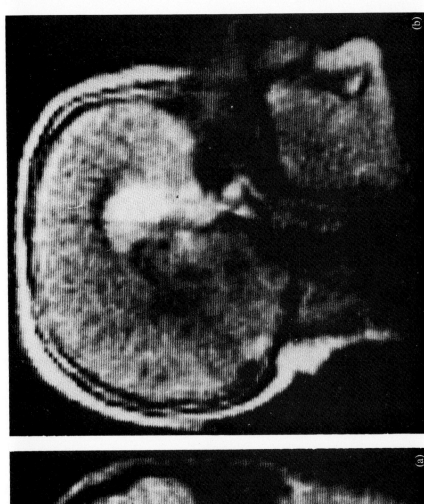

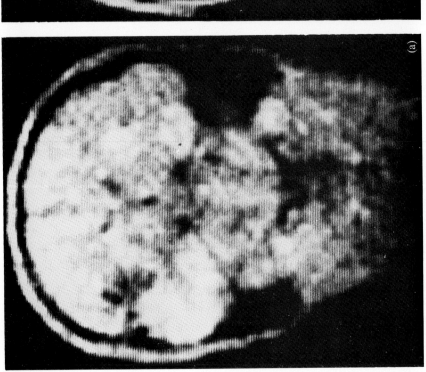

Plate 4. Head scans obtained by ^{1}H n.m.r. spin imaging that demonstrate brain pathology. (a) An arteriovenous malformation appears as a spongework of contiguous black areas. (b) A large tumour appears as a region of high signal intensity. (From Hawkes, Holland, Moore, and Worthington, 1980.)

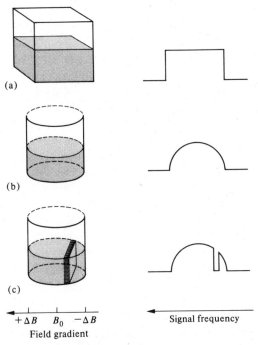

Fig. 4.6. Diagram illustrating one-dimensional imaging. (a) A rectangular beaker containing water, together with the ^{1}H n.m.r. signal obtained on applying a small field gradient along the x-axis. (b) Same as (a), except that the beaker has a circular cross-section. (c) Same as (b), except that a plastic strip containing no protons is placed across the beaker. The application of the linear field gradient generates an n.m.r. spectrum that is a one-dimensional projection of proton density along the direction of the gradient.

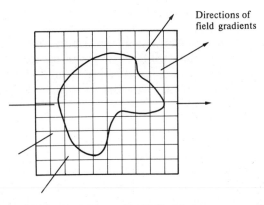

Fig. 4.7. An object, and its image, may be divided into an array of $n \times n$ picture elements. The image can be constructed by mathematical manipulation of the n signals, or projections, obtained on applying the field gradient in n different directions.

Unfortunately, the required extension to two or three dimensions is not trivial. One could apply the field gradient successively along each of three mutually perpendicular directions, but it is easy to show that the three signals that would be obtained would still be far from sufficient to provide a complete image of an object. Instead, it is necessary to apply the field gradients in a range of different directions. The number of directions that must be used depends on the spatial resolution that is required. For example, in order to obtain a two-dimensional image of an object containing $n \times n$ picture elements (see Fig. 4.7), n^2 pieces of information must be available, for otherwise there would be insufficient information to fully define the image. This means that if the signal representing each one-dimensional projection consists of n data points, the field gradient must be applied in n different directions.

This approach forms the basis of the projection—reconstruction method of n.m.r. imaging developed by Lauterbur (1973). This method has been widely used for image formation by X-ray computed tomography (X-ray CT scanning); the image is constructed by mathematical manipulation of the various signals, or projections, that are obtained. The method has been extended to three dimensions, but this introduces severe difficulties, partly because one must now solve n^3 equations in n^3 unknowns, where usually n must be greater than 100. Such a calculation can impose strains even on a modern computer. Therefore, the usual approach is to obtain a two-dimensional image of a selected thin slice of a three-dimensional object, then move on to another slice, and in this way to build up a complete three-dimensional image. Definition of a thin slice can be achieved in a number of ways; for example, by using frequency-selective radiofrequency pulses, or by means of a time-dependent field gradient (Holland, Hawkes, and Moore 1980).

Projection-reconstruction is only one of many n.m.r. imaging methods that have been devised. Many of the methods have the common feature that a linear gradient provides a one-dimensional profile, but they differ in the manner in which the second or third gradient is introduced. The methods differ in their degree of sophistication, and in the speed with which data can be acquired, but at the present moment it seems that projection—reconstruction is gaining popularity as a method that provides information reasonably rapidly and is fairly easy to implement.

An explanation of the various methods is beyond the scope of this book, but most of them were discussed at a meeting of the Royal Society of London (*Phil. Trans. R. Soc. Lond. B* 1980), and have been compared by Brunner and Ernst (1979), and by Hoult (1980 *b*).

4.2.2 Spatial resolution and imaging time

For all imaging methods, there is a strong dependence of imaging time on spatial resolution. For example, if the resolution is to be enhanced from 5 mm to 1 mm in all three dimensions, each volume element will be reduced in volume by 5^3, and will produce a correspondingly smaller signal. The time required to achieve the same signal-to-noise ratio as for the larger volume element will therefore have to be increased by a factor of $(5^3)^2$, i.e. 5^6. This sixth power dependence means that we pay heavily for enhanced resolution. If we are interested in increasing the spatial resolution in just two dimensions, within a slice of given thickness, then the imaging time will depend on (resolution)^4, which is still a severe dependence. At the present moment, published images are generated in arrays of up to 128×128 data points, regardless of the size of the object. In fact, the spatial resolution that is achieved tends to decrease as the size of the object increases; for the brain, the spatial resolution (about 2 mm) compares favourably with that of first generation X-ray computed tomography.

In view of the severe dependence of imaging time on spatial resolution, the imminent improvements in technology are likely to improve acquisition time far more significantly than spatial resolution. A reduction in imaging time is important because any bodily movement that occurs during imaging will generate artefacts in the image. For whole-body imaging, a 15-s scan time should enable an image to be made while the patient is holding his breath. It may also be possible to synchronize data collection with heart-beat, and in this way to obtain images of the heart corresponding to different phases of the cardiac cycle. In fact, this type of study has been performed on perfused hearts using ^{23}Na n.m.r. (Delayre, Ingwall, Malloy, and Fossel 1981) which can readily distinguish between the intracellular and extracellular space on the basis of their widely differing Na^+ concentrations.

4.2.3 What does the image represent?

The discussion presented above suggests that the image should represent the proton density at various points throughout an object. However, factors other than proton density can also affect the intensity of the image. In particular, the magnitudes of the relaxation times T_1 and T_2 are of importance, and under certain conditions the intensity will be approximately proportional to $\rho \, T_2/T_1$, where ρ is the proton density. For solids, $T_2 \ll T_1$, and therefore their n.m.r. response is weak. In contrast, fluids give relatively strong signals, because T_1 and T_2 are more equal. Note the similarity with conventional high-resolution n.m.r., which is far more sensitive to mobile than immobilized compounds. This dependence of image intensity on relaxation times is of great importance as a tissue contrast parameter, for tissues with the same proton density but different values of T_2/T_1 will generate distinctive responses. The image is also sensitive to diffusion, and to flow, and indeed one hope is to use n.m.r. to measure blood flow in human beings.

In general, therefore, an image will represent a mixture of parameters such as those already mentioned. However, it should often be possible to vary the experimental conditions in such a way that the image is particularly sensitive to a chosen parameter, whether this be proton density, T_2/T_1, T_1, or flow. It is interesting to note that T_1 values measured for the protons of different healthy tissues range from around 100 ms to 1 s, with relatively small standard deviations in the values obtained for each tissue type under specified conditions. Thus the highest T_1 values for normal tissue differ from the lowest by a factor of about 10. In contrast, the water concentration in soft tissues varies over a relatively narrow range; from about 69 percent in skin to 83 percent in grey brain tissue. This suggests that for certain soft tissues, discrimination and spatial resolution may be better in images portraying T_1 values than in images of proton concentration. It is also interesting that the percentage variation in T_1 values is much greater than for the properties used in other methods of medical imaging such as X-rays and ultrasound.

4.2.4 Clinical applications of spin-imaging

In 1971, Damadian noted that the T_1 values of water protons in malignant tissues of rats were much longer than the values observed in normal tissues. Since that time, there has been considerable interest in characterizing the proton signals of normal and diseased tissues, and in correlating the observed n.m.r. parameters with pathological state. Work in a number of laboratories has indeed confirmed that protons in malignant tumours of small animals have larger T_1 values than the corresponding normal tissues (see Hollis 1979). These observations opened up the possibility that n.m.r. imaging might provide an effective means of diagnosing or detecting cancer.

However, we should be cautious in considering the role of n.m.r. in cancer research, detection, and diagnosis. Firstly, some degree of care is necessary in extrapolating from the results of animal studies (many of which have been performed *in vitro* at relatively high magnetic field strengths) to results that will be obtained on human beings at the low fields used for imaging experiments. More importantly, the T_1 values measured for various normal tissues occupy a wide range between 100 ms and 1 s (Ling, Foster, and Hutchison 1980; Hollis 1979), while the T_1 values for malignant tissue can also be expected to vary over a considerable range, depending on the type of tumour and its growth rate. Therefore, there is a certain amount of overlap between normal and malignant tissue, suggesting that regrettably one cannot rely on T_1 measurements alone to distinguish conclusively between normal and malignant tissue. Finally, it has been found that there is a strong correlation between the water content of a tissue and its T_1 value, and a change in water content of just a few percent would be sufficient to account for the elevated T_1 values of malignant tissues. Therefore, *any* condition that produces oedema should generate an increase in the T_1 value of the affected region.

In summary, it is certainly possible to detect regions within humans of abnormally long T_1 values. However, in view of the points mentioned above, this would certainly not provide proof of the existence of a tumour; it would merely suggest the importance of performing further clinical investigations. This is perhaps a good illustration of the point that even if n.m.r. does prove to have wide clinical application, it should be regarded as supplementing, rather than superceding exsiting clinical methods.

One of the main advantages of n.m.r. is that repeated observations should be feasible without causing any harm to the patient. Therefore, n.m.r. imaging could be particularly invaluable in the *management* of patients with tumours; for example, it may be possible, following treatment, to distinguish residual sterile tissue from active tumour. However, it should be stressed that the detection of tumours is by no means the only clinical application of spin-imaging; a wide range of disorders associated with water concentration, diffusion, and flow should be amenable to study. For example, Hoult (1980*a*) has been developing a spin-imaging system suitable for examining premature babies, his aim being to investigate disorders such as hydrocephalus and abnormal vasculature. Similar investigations currently involve the use of X-rays, and here the lack of radiation hazard associated with n.m.r. could be particularly invaluable.

One application that could be particularly interesting is the measurement of blood flow. Some of the images that have been published take advantage of the fact that n.m.r. signals are sensitive to flow, and are able to show up the major blood vessels. However, they provide little indication of the rate of flow, quantification of which could be of great value. In principle, flow imaging is feasible, and it will be most interesting to see how this develops in future years.

4.2.5 Examples of human images

Following the demonstration that cross-sectional images could be obtained of the human wrist (Hinshaw, Bottomley, and Holland 1977), and of the forearm (Hinshaw, *et al.* 1979), images have now been obtained of the head and of the body. In plate 1 a head scan was shown alongside the expected structure. Further head scans are shown in plate 2 illustrating the point that axial transverse, coronal and sagittal views can be equally easily obtained by n.m.r. This contrasts with X-ray CT scanning, which can only generate images of other than axial transverse views by computer manipulation of the data from a large number of transverse slices. This would require a large radiation dose and a longer data acquisition time.

The n.m.r. images shown in plate 2 are each of a slice about 1 cm thick. They have a resolution within the slice of about 2 mm, and each image was acquired in about 2 min. Strong n.m.r. signals appear white. A number of general features can be pointed out. As mentioned in § 1.9.2, the lack of signal from the skull enables the surface of the brain to be seen clearly. The high signal from the exterior of the skull arises from fatty tissue within the scalp. The cerebral cortex shows up as a diffuse, peripheral white band, while the white matter deeper within the brain appears in varying shades of grey. As noted in § 1.9.2, moving liquids generate reduced signals, and so the cerebrospinal fluid within the ventricular system and the basal cisterns appears black in the images; for the same reason, blood in large vessels also gives low signal. (It should be noted that the sensitivity to motion depends on the n.m.r. excitation procedure that is used, and can be altered if so desired.) Within the orbit, retrobulbar fat generates a high signal intensity, contrasting against the globe, optic nerve, and ocular muscle. Several other features should be discernible to those familiar with the anatomy of the brain (see Hawkes, Holland, Moore, and Worthington 1980).

Few clinical tests have yet been performed, and therefore it is too soon to predict whether n.m.r. imaging will eventually become a routine clinical tool. However, two examples are given in plate 4 which show how n.m.r. images respond to pathological condition. The image of plate 4(a) reveals an arteriovenous malformation which appears as a spongework of contiguous black areas. In plate 4(b), a large tumour predictably shows up as a region of unusually high signal intensity.

Plate 3 shows some whole-body images that have been obtained recently using a method termed 'spin warp' n.m.r. imaging (Edelstein, Hutchison, Johnson, and Redpath 1980). Each volume element is 7.5 mm wide by 7.5 mm high by 18.5 mm thick (corresponding to the slab thickness), and thus has a volume of about 1 ml. The interesting feature here is that half of the images display proton density, while the others display T_1. Data collection time was about 2 min for each pair of images. The T_1 images show the more interesting detail, because the possible contrast between soft tissues is much greater if T_1 is used as the imaging parameter (see § 4.2.3). The images are not as well-resolved as those of the head, but whole body imaging is still at a relatively early stage in its development, and major improvements will undoubtedly follow. Preliminary applications to abdominal disease have now been reported using this approach to whole body n.m.r. imaging (Smith *et al.* 1981 *a, b*). In further studies, Smith, Mallard, Reid, and Hutchison (1981) have evaluated n.m.r. imaging in 30 patients with established liver disease and 20 patients without liver disease. They showed that n.m.r. easily differentiates malignant tumours from benign cystic lesions, and that the technique provides useful information in patients with cirrhosis and metastatic deposits. The images in these studies are based on measurements of proton T_1 values, rather than on proton concentration.

4.3 Magnet design

Both electromagnets and superconducting magnets are available for whole-body n.m.r. imaging. Many groups have chosen to work at the relatively low field of 0.1 T, which corresponds to a ^{1}H frequency of about 4 MHz. At this low frequency, there are no problems associated with attenuation or phase shifts of the radiofrequency fields within the sample (Bottomley and Andrew 1978). For whole body studies, such effects (see § 8.5.3) would be significant at frequencies above about 10 MHz, and could cause severe difficulties with interpretation of the images.

For experiments utilizing other nuclei, such as ^{31}P, a higher magnetic field is essential, because of the enhanced signal-to-noise and spectral resolution that this provides. Moreover, the problems associated with radiofrequency penetration are not as critical as for ^{1}H imaging, because less detail is required about spatial distribution. Thus the human forearm studies described in § 4.1.3 were performed at 32.5 MHz, using a horizontal superconducting magnet operating at a field of 1.9 T. It is likely that the same field strength will be used for the first whole body ^{31}P studies.

4.4 Safety

Before the use of whole-body n.m.r. can become routine, it is obviously obligatory to ensure that there are no significant hazards associated with the method. The three features that could provide potential hazards are (i) the radiofrequency field, (ii) the static magnetic field B_0, and (iii) the induction of electrical currents in the body resulting from rapid changes in, or movement through the field (see Budinger 1979; Hoult 1980*b*).

The heating effect associated with the radiofrequency field should be insignificant, as typical mean power levels are only a few watts, far lower than those used in diathermic therapy. In contrast to X-ray scanning, no ionizing radiation is used.

Exposure to a magnetic field of up to 2 T is very unlikely to have any ill effects, and years of experience with cyclotron magnets have provided no firm evidence of any hazard. Rapid movement within the static field, or alternatively the use of rapidly switched field gradients can cause electrical currents to be induced within the body. Although any effects are likely to be very small, this is potentially the most harmful aspect of n.m.r., and clearly further safety tests are called for, in order to check all of the possible hazards. However, it seems probable that whole body n.m.r., performed under controlled clinical conditions, will be free from hazard, and indeed will be much safer than the majority of the imaging methods that are currently in use.

5

The theoretical basis of the n.m.r. experiment

5.1. Introduction

In order to understand how radiation is absorbed by matter, we need to recognize firstly that radiation is quantized and secondly that atoms and molecules can only have certain discrete energy levels. These are concepts that are outside the realm of classical physics, and therefore a rigorous treatment of any spectroscopic technique requires the use of a branch of physics that is termed quantum mechanics. However, certain aspects of n.m.r. can be understood surprisingly adequately and clearly in terms of classical physics. For this reason descriptions of n.m.r. theory frequently contain a mixture of quantum-mechanical and classical treatments, depending on which provides the simpler or more helpful picture of what is happening. In this chapter, we shall first give a brief quantum mechanical description of the basic properties of atomic nuclei, explaining how magnetic nuclei interact with applied magnetic fields. The treatment that is presented here is intended to be comprehensible to readers with no prior knowledge of quantum mechanics; regrettably, however, some of the results that are given will have to taken on trust, for the theory on which they are based is complicated and far beyond the scope of a book of this type. We shall then proceed to describe n.m.r. in terms of the classical model that is particularly well suited to explaining many of the practical aspects of n.m.r., such as the effects of radiofrequency pulses. We also discuss what is meant by the Fourier transform, and explain its critical role in the n.m.r. experiment.

We begin by noting that certain atomic nuclei, such as ^1H, ^{13}C, and ^{31}P, possess a property known as spin, which can be visualized as a spinning motion of the nucleus about its own axis. These nuclei, in common with other spinning objects, possess the property of angular momentum, and they also, by virtue of their electrical charge, have magnetic properties. This nuclear magnetism is analogous to the magnetism generated by an electrical current circulating in a small loop of wire; such a current loop behaves like a small bar magnet, and similarly the charged, spinning nucleus can be regarded as a tiny bar magnet, rotating about its own axis (see Fig. 1.3(a)).

The expression 'magnetic dipole moment' (or more simply 'magnetic moment') is often used to describe the properties of magnetic objects; it defines the turning moment that is experienced by the object when it is placed in an applied magnetic field, and this is the term that is used to describe the properties of magnetic nuclei. As we shall see, the angular momentum and magnetic moment are closely related to each other, and it is for this reason that we start the quantum mechanical section with a brief discussion of the quantised nature of the nuclear angular momentum.

5.2. Quantum mechanical description

5.2.1. The atomic nucleus

Quantum mechanics tells us that the angular momentum of atomic nuclei can only have certain discrete values, specified by a quantum number I. The magnitude p of the nuclear angular momentum is given by

$$p = \hbar\{I(I + 1)\}^{1/2},\tag{5.1}$$

where $\hbar$ is equal to $h/2\pi$, h being the Planck constant. The quantum number I, usually called the spin of the nucleus, may only have integral or half-integral values as follows:

(i) I is integral for nuclei with even mass number.

(ii) I is zero for nuclei with even numbers of both neutrons and protons. This is because of the tendency for both neutrons and protons to form pairs in such a way that the individual spins cancel out. Therefore, ^{12}C and ^{16}O have zero spin and do not produce n.m.r. signals.

(iii) I is half-integral for nuclei with odd mass numbers. Nuclei of spin $\frac{1}{2}$, such as ^{1}H, ^{13}C, and ^{31}P, are particularly important in n.m.r. as these are the nuclei that give rise to high-resolution spectra.

Angular momentum is a vector property (see Appendix 5.1), and is specified by both magnitude and direction. In order to specify the direction of the angular momentum, it is necessary to introduce a second quantum number m, for it is found that the angular momentum vector can only have certain discrete orientations with respect to any given direction (e.g. the z-direction). The component p_z of angular momentum along the z-direction is given by

$$p_z = m\hbar.\tag{5.2}$$

m may have any of the $2I + 1$ values, $I, I-1, \ldots -I$, and so for a nucleus of spin

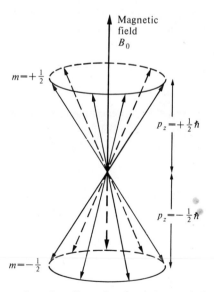

Fig. 5.1. Diagram illustrating the allowed orientations of the angular momentum vector of a nucleus of spin $\frac{1}{2}$. The orientations are specified by the quantum number m, and describe two cones.

$\frac{1}{2}$, m can be $+\frac{1}{2}$ or $-\frac{1}{2}$. Therefore, for such a nucleus

$$p_z = \pm\tfrac{1}{2}\hbar, \qquad (5.3)$$

and these two spin states are illustrated in Fig. 5.1.

As mentioned in the introduction, the magnetic moment of a nucleus is closely related to its angular momentum; in fact the magnetic moment has the same direction as the angular momentum, and has magnitude μ given by

$$\mu = \gamma p, \qquad (5.4)$$

where γ is a proportionality constant known as the magnetogyric (or gyromagnetic) ratio of the nucleus. Thus the component of the magnetic dipole moment along the z-axis can be written

$$\mu_z = \gamma\hbar m. \qquad (5.5)$$

It should be noted that the magnitude of γ, and hence of the nuclear magnetic moment, cannot be predicted from classical physics. For most nuclei γ is positive, but for some others, such as ^{15}N, γ is negative.

5.2.2. The interaction of the nucleus with a static magnetic field

If a static magnetic field[1] B_0 is applied along the z-axis, the nucleus acquires energy E as a result of the interaction between the field B_0 and the nuclear moment, the magnitude of which is given by

$$E = -\mu_z B_0$$

or, from eqn (5.5)

$$E = -\gamma\hbar m B_0. \qquad (5.6)$$

Since m can have any of the values $I, I-1, \ldots, -I$, the nuclear energy levels are split into $2I + 1$ states by the application of the field, as was shown in Fig. 1.4 for a nucleus of spin $I = \frac{1}{2}$. The energy difference ΔE between adjacent states is given by

$$\Delta E = \gamma\hbar B_0. \qquad (5.7)$$

5.2.3. The effect of an oscillating magnetic field

In the n.m.r. experiment transitions between adjacent states are induced by application of a suitable oscillating magnetic field B_1 in the xy-plane. The field must oscillate with a frequency ν_0 given by

$$\Delta E = h\nu_0.$$

Therefore, using eqn (5.7) we find that

$$\nu_0 = \gamma B_0/2\pi \qquad (5.8)$$

If we write $\omega_0 = 2\pi\nu_0$, where ω_0 is known as the angular frequency, then

$$\omega_0 = \gamma B_0. \qquad (5.9)$$

---

[1] A magnetic field can be described in terms of its flux density B_0 which determines the force exerted on other magnets. In this book we refer to B_0 simply as the magnetic field rather than the magnetic flux density which is a somewhat unwieldy and unfamiliar expression.

Equation (5.8), or eqn (5.9), expresses the resonance condition for n.m.r. experiments. Note that only transitions between adjacent states take place. This is an example of quantum-mechanical 'selection rules' that govern transitions between energy levels. Since γ differs for each nuclear isotope, different nuclei resonate in a given field B_0 at widely different frequencies. At conventional values of B_0 the frequencies occur in a convenient radiofrequency band, and the oscillating field B_1 is commonly referred to as the radiofrequency (r.f.) field.

5.2.4. The populations of the nuclear energy states

The populations of nuclei in the various energy states are determined by the Boltzmann distribution, i.e. at a thermal equilibrium characteristic of the temperature T the relative numbers n^+ and n^- of nuclei in the spin $+\frac{1}{2}$ and $-\frac{1}{2}$ states are given by

$$n^-/n^+ = \exp(-\Delta E/kT)$$
$$= \exp(-\gamma\hbar B_0/kT) \qquad (5.10)$$

where k is the Boltzmann constant (see Fig. 5.2). For protons in a magnetic field of 5 T the magnitude of ΔE is only 10^{-6} eV whereas at $20\,^{\circ}$C the thermal energy kT is about 2.5×10^{-2} eV. Therefore $\exp(-\Delta E/kT)$ is very close to unity, and the populations differ by less than 1 part in 10^4. Absorption of energy by the nuclei from the oscillating magnetic field B_1 relies on there being a population difference between adjacent states; if the populations were equal there would be equal numbers of transitions in both directions, resulting in no net absorption of energy and no signal. The small energy difference between the states therefore leads to a very weak absorption of energy and is responsible for the inherently low sensitivity of n.m.r. Note that an increase in the magnetic field B_0 increases the energy difference between adjacent states (eqn (5.7)) and hence their population difference (eqn (5.10)), and therefore considerably enhances the net absorption of energy. As a result, a high magnetic field is generally desirable to improve the signal-to-noise ratio.

There are two aspects of this basic quantum-mechanical treatment that are consistent with the adequacy of the classical picture of n.m.r. Firstly, the Planck

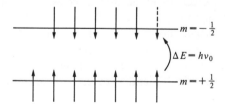

Fig. 5.2. Nuclei of spin $I = \frac{1}{2}$ in their two possible spin states. The nuclei of spin $m = +\frac{1}{2}$ (upward-pointing arrows) have lower energy than those of spin $m = -\frac{1}{2}$ (downward-pointing arrows); the $m = +\frac{1}{2}$ state is therefore slightly the more populated of the two states. A transition from the lower to the higher state is indicated by the curved arrow; transitions such as this tend to equalize the populations and hence lead to saturation (see § 5.5.3).

constant is absent from eqn (5.9). Secondly, the predictions of quantum theory and classical physics differ considerably for systems in which the separation ΔE of the energy levels is much greater than the mean thermal energy kT. However, they tend to converge when ΔE is much less than kT, which is true in the case of n.m.r. We now move on to the classical description of n.m.r. to continue our picture of how n.m.r. signals are excited and detected.

5.3. Classical description of n.m.r.

If a bar magnet is placed in a magnetic field it aligns itself parallel to the field, because this is the orientation of least energy. However, if the magnet possesses angular momentum it will not align parallel to the field, but instead will precess about the field with a characteristic angular frequency ω_0 (see Fig. 5.3). (Angular

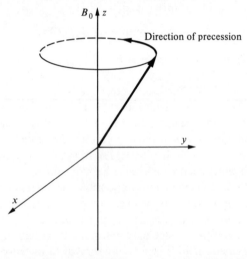

Fig. 5.3. A magnet possessing angular momentum precesses in an applied field B_0. The heavy arrow indicates the orientation of the magnet, and of its angular momentum.

frequency is the rate of rotation in radians per second.) This behaviour is analogous to that of a spinning gyroscope in the earth's gravitational field. If we regard the nucleus as a tiny bar magnet possessing angular momentum, it is possible to show, using classical physics, that the nuclear magnet, or moment, precesses about the direction of an applied field B_0 with angular frequency $\omega_0 = \gamma B_0$. This is known as the Larmor frequency and is identical to the resonance frequency derived from quantum theory (see eqn (5.9)).

In n.m.r. experiments we do not of course study individual nuclei, but rather a sample containing a large number (typically about 10^{18}) of magnetic nuclei. These nuclear magnets all precess about B_0 in the same direction, regardless of the value of their quantum number m. Since there is no preferred orientation in the plane perpendicular to B_0 (the xy-plane), the net component of magnetic moment in the xy-plane is zero. However, there is a net magnetization (defined as magnetic

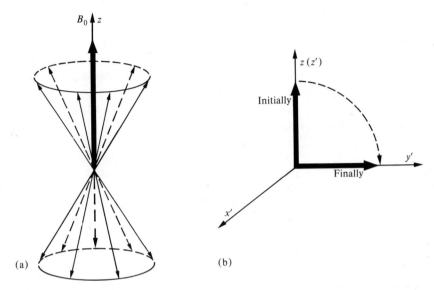

Fig. 5.4. (a) Prior to the application of a radiofrequency pulse, the *net* component of magnetisation, indicated by the heavy arrow, is along the z-axis. (b) The effect of a $90°$ pulse applied along the x'-axis is to tilt the magnetisation on to the y'-axis. (The z-axis of the laboratory frame and the z'-axis of the rotating frame are equivalent.)

moment per unit volume) along the z-axis (see Fig. 5.4(a)) because there are slightly more nuclei oriented with the field than against it, i.e. there are more nuclei in the lower energy state than in the upper state.

In order to detect the magnetization set up within a sample by the B_0 field, we shall see that it is necessary to tilt the magnetization towards or into the xy-plane. It turns out that this can be accomplished by means of a radiofrequency field applied in the xy-plane. In order to understand how this process works it is most useful to appreciate the concept of the rotating frame of reference, which is perhaps not as awesome as it sounds.

5.3.1. The rotating frame of reference and the B_1 field

We have seen that the application of a static field B_0 causes the nuclear magnets to precess about the direction of B_0 with angular frequency $\omega_0 = \gamma B_0$. Let us consider a world rotating with angular frequency ω_r relative to the laboratory. In this so-called rotating frame of reference the nuclear magnets appear to precess with angular frequency $\omega_0 - \omega_r$. If ω_r happens to be equal to ω_0, then in this particular frame of reference $\omega_0 - \omega_r$ is equal to zero, i.e. the nuclear magnets do not appear to precess at all, and by analogy with eqn (5.9) the apparent static magnetic field must also be zero. We shall now imagine ourselves to be in this rotating frame of reference for which $\omega_0 = \omega_r$. By analogy with the laboratory frame, which has coordinates x, y, and z, we label the coordinates of the rotating frame x', y', and z' (in fact z and z' are equivalent).

Let us apply along the x'-axis of the rotating frame of reference a field B_1 which is static in this frame. Then *just as the application of B_0 causes the nuclei to precess about its direction with angular frequency γB_0, so the application of B_1 in the rotating frame causes the nuclear magnetization to precess about B_1 with angular frequency γB_1*. This is because the B_1 field is the only apparent field experienced in the rotating frame. If the field B_1 is applied for time t_p, the nuclei will rotate through an angle $\theta = \gamma B_1 t_p$. The nuclei will rotate through an angle of $90°$ (corresponding to $\theta = \pi/2$ rad) if t_p is such that

$$\gamma B_1 t_p = \pi/2. \tag{5.11}$$

A pulse of B_1 field that has this duration is known as a $90°$ pulse, and its effect is to tilt the net magnetization away from the z- (or z'-) axis into the $x'y'$-plane of the rotating frame (see Fig. 5.4(b)). (Similarly, a $180°$ pulse tilts the magnetization on to the negative z-axis.)

Let us now consider the nature of the field B_1 which is static in the rotating frame. If we revert to the normal laboratory frame, we see that B_1 must correspond in this frame to a field that rotates about the z-axis with angular frequency ω_0. The simplest method of generating this rotating B_1 field is to apply an oscillating field along a given direction (e.g. the x-direction) in the xy-plane, for such a field can be broken down into two components rotating in opposite directions. This is exactly equivalent to the breaking down of plane polarized light into left- and right-handed circularly polarized light. The component rotating in the opposite direction to the nuclear precession can be shown to have a negligible effect in n.m.r., and so the oscillating B_1 field is effectively equivalent to the required rotating field.

We therefore find that macroscopic changes in the nuclear magnetization can be induced by applying an oscillating B_1 field that has the same frequency as the Larmor frequency of the nuclei. This classical result agrees perfectly with the quantum-mechanical result which states that resonance is achieved by application of magnetic radiation of frequency ω_0 in the xy-plane. The equality of the Larmor frequency and the frequency of the B_1 field is analogous to other resonance phenomena for which the 'driving' frequency is equal to a characteristic frequency of the system. The B_1 field is in fact generated simply by passage of an oscillating electric current through the transmitting coil that surrounds the sample.

5.4. The basis of signal detection

In the absence of an applied B_1 field the nuclear magnets precess randomly about the field B_0 at their characteristic Larmor frequency. At any instant the net component of magnetization in any direction within the xy-plane is zero and no signal is observed. Now consider the effect of a pulse of radiofrequency field B_1 which tilts the net nuclear magnetization away from the z-axis towards the $x'y'$-plane of the rotating frame, and therefore towards the xy-plane of the laboratory frame. Following the pulse, the nuclear spins experience only the static field B_0 and so they continue to precess about B_0. However, now the phase of their precession is *not* totally random for a net component of magnetization M_{xy} has been generated in the xy-plane. This component M_{xy} rotates *coherently* about B_0 at the frequency

ω_0 and induces an electromotive force (e.m.f.) in the coil surrounding the sample. The induced e.m.f. oscillates at the frequency ω_0 and its magnitude is governed by Faraday's law of magnetic induction. This e.m.f. is then amplified and processed on the receiving side of the n.m.r. spectrometer to give a recognizable n.m.r. signal. Note that the same coil can be used for transmitting the B_1 field and for detecting the resulting signal. The radiofrequency coil is a critical part of the n.m.r. spectrometer, and is discussed in detail in Chapter 8.

5.5. Fourier transform n.m.r.

In § 1.7.2 we briefly discussed continuous-wave and Fourier transform n.m.r., which are the two main modes of detection of n.m.r. signals. In the continuous-wave mode the B_1 field is applied continuously in the xy plane, and the B_0 field is swept through a range of field strengths in order to obtain a spectrum in which signal amplitude is plotted as a function of field strength. This method was used until the 1970s for the observation of high-resolution n.m.r. spectra, but it has gradually been replaced, particularly for biological studies, by the relatively new technique of Fourier transform n.m.r.

The main disadvantage of continuous-wave n.m.r. stems from the fact that at any given time only one resonance is being detected. The accumulation of a high-resolution spectrum is therefore an inefficient process in that only a fraction of the total collection time is spent observing any single resonance. It would be very advantageous if a technique could be developed whereby all the resonances could be detected simultaneously. The uses of pulses of radiofrequency field were recognized from the early days of n.m.r., but pulsed n.m.r. was little used in chemical analysis for many years. Then in 1966 Ernst and Anderson showed that by use of Fourier transformation the pulsed approach could give high-resolution spectra equivalent to those obtained in continuous-wave n.m.r. Pulsed n.m.r. has the great advantage over continuous-wave n.m.r. that all the resonances can be excited simultaneously, and now almost all high-resolution biological studies are performed on Fourier transform n.m.r. spectrometers.

5.5.1. The Fourier transform

In Fourier transform n.m.r. the signal is collected as a function of time following the application of a radiofrequency pulse of B_1 field. In order to obtain a conventional n.m.r. spectrum in which signal amplitude is plotted as a function of frequency, it is necessary to manipulate the response to the radiofrequency pulse in the appropriate manner. This manipulation is performed by means of the mathematical device known as Fourier transformation which enables a quantity that varies with time to be analysed in terms of the sum of its frequency components. In this section we shall present a qualitative pictorial consideration of the way in which signals can be represented as a function of either time or frequency. In the next section we discuss the important concept of phase, while in Appendix 5.3 a more mathematical description of the Fourier transform is given.

Let us consider the analogy of a tuning fork. When it is struck it responds by ringing with a characteristic frequency. This ringing can be regarded as a wave of

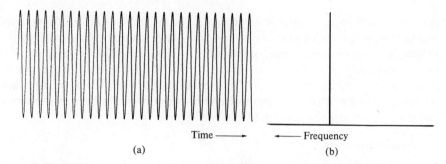

Time ⟶   ⟵ Frequency

(a)   (b)

Fig. 5.5. A wave can be expressed as a function of time or of frequency.

the form shown in Fig. 5.5(a). Alternatively, the sound can be expressed in terms of its frequency as shown in Fig. 5.5(b) where the amplitude of the response is plotted as a function of frequency. Figure 5.5(b) expresses the perfect quality of the note in the sense that it contains one single-frequency component. A single-frequency component corresponds to a perfect sine or cosine wave which continues for ever, and so the long lifetime of the response as shown in Fig. 5.5(a) equally expresses the perfect quality of the note.

Suppose that the note is less ideal in that it contains a distribution of frequencies centred about ν_0 as shown in Fig. 5.6(b). Let us now consider the nature of the

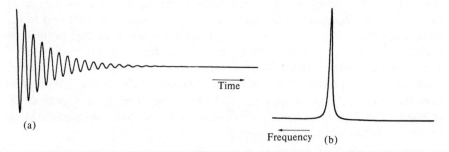

Time

(a)

Frequency   (b)

Fig. 5.6. A waveform expressed as a function of time and of frequency.

ringing. Initially, the response has a large amplitude which represents the fact that all frequency components start off in phase with each other. However, because the various components all oscillate at slightly different frequencies, they will gradually become out of phase with each other and destructive interference will take place. The amplitude of the ringing will gradually decay to zero as shown in Fig. 5.6(a). If the frequency distribution were greater, the decay would take place more rapidly; there is an inverse relationship between the length of time for which the ringing takes place and the spread in frequency. The two representations shown in Fig. 5.6 (and in Fig. 5.5) are equivalent[1]; the one specifies the other and they are connected

[1] Provided that phase information is preserved in the frequency response (see § 5.5.2 for a discussion of phase).

to each other by the mathematical device of Fourier transformation. If the decay of the ringing is exponential, then its time constant T_c is related to the width at half-height of the frequency response by the relationship

$$1/T_c = \pi \Delta \nu_{1/2} \qquad (5.12)$$

The similarity between this equation and eqn (1.8) should be noted.

If several different tuning forks were struck simultaneously, the response expressed as a function of time would be complicated, as shown in Fig. 5.7(a). Nevertheless, the human ear would be capable of discriminating between the constituent frequencies, and indeed a musical person could identify the precise frequencies that were present. Effectively, he would be Fourier transforming a complicated wave-pattern to give the frequency distribution shown in Fig. 5.7(b).

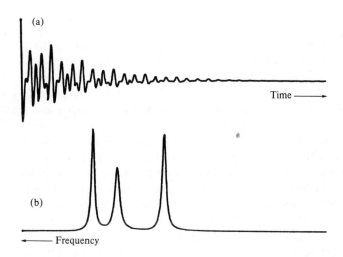

Fig. 5.7. A waveform expressed as a function of time and of frequency.

Now consider a wave of the form shown in Fig. 5.8(a). This is a wave of frequency ν_0 which is cut off after a time t. Suppose we wish to measure the frequency. We could imagine a device that does this by counting the number of maxima in the waveform in the time t. In the example shown, where t is 1 s, the frequency would be measured to be 11 waves per second, i.e. 11 Hz. However, our device might equally measure the number of minima, in which case the frequency would appear to be 10 Hz. There is an inherent uncertainty in the measured frequency which corresponds to approximately one wave in the time interval of the measurement, and this is true regardless of the duration of the time interval. Therefore, if the frequency of a waveform is measured over a time period t, there is necessarily an uncertainty in the measured frequency of 1 wave in t seconds, i.e. $1/t$ Hz. This is equivalent to there being an effective spread in frequency $\Delta \nu \approx 1/t$.

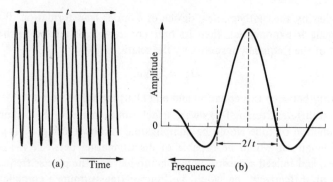

Fig. 5.8. (a) A pulse of radiation, and (b) its frequency distribution. This frequency distribution differs slightly from the frequency *response* to the pulse of an n.m.r. system.

Such a result can be obtained rigorously by Fourier analysis, and the precise frequency distribution corresponding to the pulse is shown in Fig. 5.8(b). The expression $\Delta\nu \approx 1/t$ is equivalent to one of the uncertainty principle relationships of quantum mechanics and is similar in form to eqn (5.12). The physical meaning of Fig. 5.8 is that the pulse shown in Fig. 5.8(a) could be generated by summing a distribution of waves of infinite duration whose frequencies and amplitudes are given by Fig. 5.8(b). Since $\Delta\nu \approx 1/t$, the frequency spread increases as t decreases.

Let us now return to the detection of n.m.r. signals. We wish to optimize the sensitivity of the technique by simultaneously exciting all of the nuclei, the resonance frequencies of which may extend over a range of, say, 5 kHz. We now see that we should be able to do this by applying a pulse of radiofrequency field, the duration of which is sufficiently short for the effective frequency distribution $\Delta\nu$ to be much greater than 5 kHz. Therefore, using $\Delta\nu \approx 1/t$, the duration of the pulse must be much less than $200\,\mu s$[1]. Thus in Fourier transform n.m.r. the radiofrequency field is applied in the form of short (typically about $20\,\mu s$) pulses of sufficient power to ensure that during this short time the magnetization is tilted through a reasonably large angle.

The response of the nuclei to the pulse of the radiofrequency field B_1 is totally analogous to the ringing of the tuning forks discussed above. Following the application of the pulse, the net magnetization has a component M_{xy} in the xy-plane. When the pulse is switched off the nuclei precess freely at their own characteristic Larmor frequencies about the field B_0. The coherence of the precession gradually disappears owing to the inherent frequency distribution resulting from spin—spin relaxation processes, and the magnetization M_{xy} decays with its characteristic time constant T_2. If additional factors such as magnetic field inhomogeneity contribute

[1] In fact, although this argument holds qualitatively, in quantitative terms the precise frequency *response* of the n.m.r. system differs slightly from the frequency distribution of the *applied* pulse. The effects of finite pulse width are discussed in more detail in § 7.2.5. A corollary of this argument is that if we wish to apply a selective pulse that excites signals from only a narrow frequency band, we can do this by using a long weak pulse of the appropriate frequency for a long pulse has only a small frequency distribution.

to line broadening, the decay is even more rapid. The observed decay is thus characterized by the time constant T_2^* where $T_2^* \leqslant T_2$. The expression 'free-induction decay' (FID) is often used to describe the signal resulting from a radiofrequency pulse; it describes the *decay* of the *induced* signal arising from *free* precession of the nuclei in the field B_0.

If nuclei are present in several different environments, the resulting FID contains contributions at several different frequencies, each decaying with its own characteristic time constant. Fig. 5.7(a), which shows the ringing of three tuning forks, could equally represent the FID from nuclei in three different chemical environments. The Fourier transform of this decay would then give the conventional n.m.r. spectrum shown in Fig. 5.7(b).

5.5.2. The concept of phase

In this section, we introduce the concept of phase, which is tremendously important in the treatment of waveforms and turns up repeatedly in n.m.r. (see §§ 7.3 and 7.4). Consider magnetization M_{xy} rotating about the z-axis with angular frequency ω_0 as shown in Fig. 5.9. If M_{xy} is oriented along the x-axis at time $t = 0$,

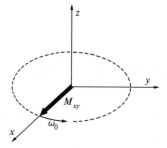

Fig. 5.9. Magnetization M_{xy} rotating about the z-axis with angular frequency ω_0.

its rotational motion can be represented by the expressions

$$M_x = M_{xy} \cos \omega_0 t$$

$$M_y = M_{xy} \sin \omega_0 t$$

where at any given time M_x and M_y are the components of magnetization along the x- and y-axes. Note that M_y can be written

$$M_y = M_{xy} \cos(\omega_0 t - \phi)$$

where ϕ (which in this case is equal to $90°$ or $\pi/2$ rad) is called the phase angle and represents the difference in phase between M_x and M_y. M_x and M_y differ only in that one is retarded relative to the other by a quarter of a cycle; the corresponding phase difference is $90°$.

In an n.m.r. experiment M_{xy} following a radiofrequency pulse decays with a time constant T_2^*, so we could write

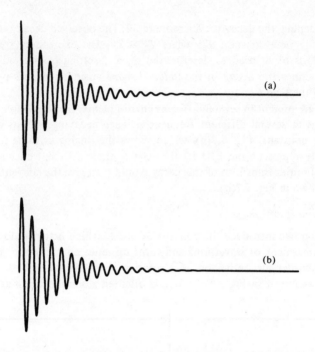

Fig. 5.10. Two free-induction decays, 90° out of phase with each other. Note that in (a), the signal has its maximum value at zero time, whereas in (b) the signal is zero at zero time.

$$M_x = M_{xy} \cos \omega_0 t \exp(-t/T_2^*)$$

$$M_y = M_{xy} \sin \omega_0 t \exp(-t/T_2^*).$$

The FID shown in Fig. 5.10(a) represents M_x, while the decay shown in Fig. 5.10(b) represents M_y. Fourier transformation of either of these waveforms gives two components as shown in Figs. 5.11 and Fig. 5.12. These components are termed real and imaginary, for reasons discussed in Appendices 5.2. and 5.3. The form shown in Fig. 5.11(a) has the characteristic Lorentzian line shape $g(\omega)$ given by

$$g(\omega) \propto \frac{T_2^*}{1 + T_2^{*2} (\omega - \omega_0)^2}.$$

This is the form that characterizes the absorption of energy by the nuclear spins. It is termed the absorption mode, and it is in this form that the spectrum is finally displayed. The form shown in Fig. 5.11(b) is commonly called the dispersion mode for reasons given in Appendix 5.3. The difference between the transforms shown in Figs. 5.11 and 5.12 reflects the fact that the free-induction decays that are being transformed are 90° out of phase with each other. Figure 5.11 can therefore be converted into Figure 5.12 by applying a 'phase correction' of 90° to the transformed spectra. This would be equivalent to changing the phase of the corresponding FID by 90°.

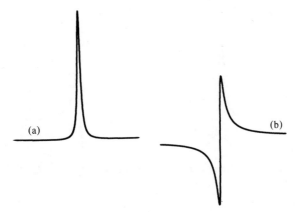

Fig. 5.11. The Fourier transform of the decay shown in Fig. 5.10(a). (a) is the real component, and (b) is the imaginary component.

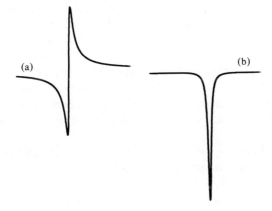

Fig. 5.12. The Fourier transform of the decay shown in Fig. 5.10(b). (a) is the real component, and (b) is the imaginary component.

It can be shown that the observed free-induction decay is in fact a measure of the component of magnetization along a given axis of the *rotating* frame of reference. The precise axis is determined by the way in which the spectrometer is set up and is somewhat arbitrary; hence the phase of the decay and of its resulting Fourier transform is also arbitrary. As a result, the real and imaginary components of the Fourier transform might be of the form shown in Fig. 5.13. Conversion to the required form of Fig. 5.11 is accomplished by applying the appropriate phase correction to the transformed spectrum. Following phase correction, the required spectrum is present in the first half (i.e. the 'real' part) of the computer memory and can be displayed or plotted out as desired.

It should be noted that most modern spectrometers now accumulate two FIDs 90° out of phase with each other. This is known as quadrature detection and has significant advantages, as discussed in § 7.3.2.

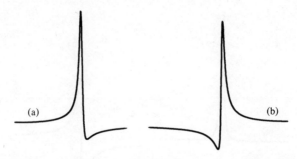

Fig. 5.13. The real and imaginary components of an arbitrary Fourier transform, prior to phase correction.

5.5.3. Accumulation of spectra – T_1 and the effects of saturation

For most biological samples the signal obtained on Fourier transformation of a single FID is too weak to be clearly distinguishable from noise. It is therefore almost always necessary to improve the signal-to-noise ratio by adding together a large number of FIDs. The accumulation of N FIDs leads to an improvement of $\sqrt{N}$, because the signal increases by a factor of N whereas the noise, being random, increases only by $\sqrt{N}$. Thus in Fourier transform n.m.r., a radiofrequency pulse is applied, the FID is observed, and the process is repeated at chosen time intervals until a sufficient signal-to-noise ratio is obtained. Fourier transformation of the data is then performed using an on-line computer to produce a conventional n.m.r. spectrum.

The choice of time interval between consecutive radiofrequency pulses is affected by a process known as saturation. Consider as an example the effect of a 90° radiofrequency pulse (see § 5.3.1). The pulse tilts the magnetization away from the z-axis into the xy-plane, and we detect the resulting magnetization M_{xy} that is developed in this plane. M_{xy} decays with a time constant T_2^* and therefore almost disappears after a time of about $4T_2^*$. It might seem appropriate to apply the next radiofrequency pulse at this time, but this is not necessarily the best procedure for optimizing the signal-to-noise ratio.

The effect of each 90° pulse is to 'sample' the z-component of magnetization M_z that exists immediately prior to the pulse. Just after the first 90° pulse M_z is zero, and if M_z were still zero after a time $4T_2^*$ a second 90° pulse applied at this time would produce no further signal. Such a situation is known as saturation. Fortunately, mechanisms exist whereby M_z returns to the equilibrium value that existed prior to the application of the first radiofrequency pulse. This return to equilibrium is known as relaxation; it is often an exponential process and is characterized by the time constant T_1 which is known as the spin–lattice (or longitudinal) relaxation time (see Fig. 5.14).

For biological samples T_1 is usually much longer than T_2^*. As a result, after time $4T_2^*$, M_z has recovered very little, and it is certainly not optimal to apply each 90° pulse immediately following the previous FID. Conversely, it is time wasting to

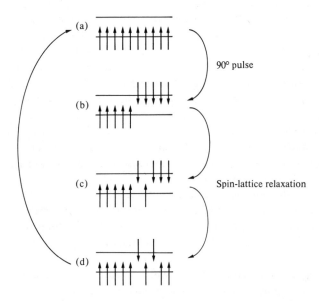

Fig. 5.14. Schematic diagram illustrating the effects of a 90° pulse and of spin-lattice relaxation. (a) At equilibrium, there is *net* magnetization in the z-direction, because there are more nuclei in the low than in the high energy state. This net magnetization M_z is represented in this figure by 10 arrows (nuclear magnets) pointing upwards. (b) Immediately after a 90° pulse, the net magnetization in the z-direction is zero. (There is, however, magnetization in the xy-plane, which is not illustrated in this figure.) (c) and (d) The magnetisation M_z gradually returns to its equilibrium value, represented by (a), with a time constant T_1.

wait until M_z has almost completely recovered to its equilibrium value. If we wish to optimize the signal-to-noise ratio obtainable in a given time a compromise is necessary, and it is found that the optimal time interval between consecutive 90° pulses is $1.25T_1$. However, it is even better to apply smaller-angle pulses rather more rapidly; this is discussed further in § 7.2.2.

5.6. A comparison of continuous-wave and Fourier transform n.m.r.

The main advantage of Fourier transform n.m.r. over continuous-wave n.m.r. is that it can provide a far higher signal-to-noise ratio over a given period of time. The difference between the two methods arises because in Fourier transform n.m.r. all the resonances of a spectrum are excited simultaneously, whereas in continuous-wave n.m.r. they are excited one by one. An estimate of the sensitivity enhancement provided by Fourier transform n.m.r. can be obtained from the following argument, which admittedly contains some simplifying assumptions.

For a spectrum containing a single resonance, the signal-to-noise ratio obtained over a given period of time will be similar for the two methods, as for each method the signal and the noise at the signal frequency are observed over the same length of time. Now consider a spectrum of total bandwidth w which contains a large number of resonances each of linewidth $\Delta\nu_{1/2}$. When continuous-wave n.m.r. is used each resonance will be detected for only a fraction $\Delta\nu_{1/2}/w$ of the total accumulation time. In contrast, in Fourier transform n.m.r. each resonance is detected over the whole duration of the experiment (if $T_1 \approx T_2^*$). Therefore, the continuous-wave experiment must take $w/\Delta\nu_{1/2}$ times as long as the Fourier transform experiment to acquire the same signal-to-noise ratio. Alternatively, since the signal-to-noise ratio increases as the square root of the accumulation time, Fourier transform n.m.r. provides an improvement in the signal-to-noise ratio of $\sqrt{(w/\Delta\nu_{1/2})}$ over a given period of time. This improvement will be rather less for samples in which T_1 is much longer than T_2^*. The difference in sensitivity between continuous-wave and Fourier transform n.m.r. is discussed in detail by Ernst and Anderson (1966), Ernst (1966), and Shaw (1976).

A second distinction between the two detection methods concerns the effects of saturation. In Fourier transform n.m.r. saturation reduces signal intensities but has no effect on lineshapes. In contrast, the effects of saturation in continuous-wave n.m.r. are to distort and broaden lineshapes in addition to reducing signal intensities.

There is a third detection method which is similar to continuous-wave n.m.r. except that it employs a rapid sweep through the spectra. Known as correlation spectroscopy, it was developed by Dadok and Sprecher (1974) and can provide a similar sensitivity to Fourier transform n.m.r. In practice, the method is particularly useful as a means of avoiding the large solvent signal in ^1H n.mr., but as yet it has not had widespread use.

Appendix 5.1. Vectors

Scalar quantities such as temperature, mass, and length are specified by magnitude alone. However, some physical quantities are only completely specified if their direction is given in addition to their magnitude. Such quantities are known as vectors. Magnetic field, angular momentum, and magnetic moment are all vector quantities; angular momentum, together with other quantities that describe rotational motion, is specified by a vector the direction of which gives the axis about which rotation takes place (see under vector multiplication below). Vector algebra therefore figures prominently in basic n.m.r. theory, and it should be useful to discuss some simple aspects of the subject.

Vector addition and subtraction

A vector can be represented diagramatically by an arrow pointing along the direction of the vector and of length equal to its magnitude. The addition of two vectors

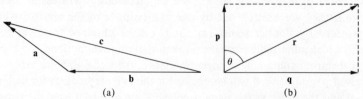

Fig. 5.15. (a) Vector diagram in which **c** is the resultant of **a** and **b**. (b) Vector diagram in which **r** is resolved into two perpendicular components **p** and **q**.

is performed using the triangle law of addition as illustrated in Fig. 5.15(a). The vector **c** is the sum or resultant of the two vectors **a** and **b**. The difference **a** − **b** is found in a similar manner, except that the direction of **b** is reversed.

Any vector **r** can be resolved into two components at right-angles; this is effectively the converse of vector addition. Figure 5.15(b) shows how **r** is resolved into two perpendicular components **p** and **q**. The component of **r** along the direction of **p** is given by the magnitude of **p**, and is equal to $r \cos \theta$. Similarly, the component of **r** along **q** is $r \sin \theta$. Any vector **r** can be resolved in a similar manner into three components parallel to the x, y, and z-axes of a Cartesian co-ordinate system.

Vector multiplication

Multiplication of a vector quantity by a scalar quantity s simply involves multiplying the magnitude of the vector by the factor s and has no effect on the direction of the vector. Thus **r** multiplied by s is equal to s**r**.

It is often convenient to introduce the unit vectors **i**, **j**, and **k** as vectors of unit magnitude that are parallel to the x-, y-, and z-axes respectively of a Cartesian coordinate system. A vector **r** can then be written

$$\mathbf{r} = a\mathbf{i} + b\mathbf{j} + c\mathbf{k}$$

where a, b, and c are the magnitudes of the components of **r** parallel to the three axes.

The *scalar product* of two vectors **p** and **q** is written **p** · **q**, and is sometimes called the *dot product*. It is a scalar quantity of magnitude $pq \cos \theta$, where θ is the angle between **p** and **q**. The scalar product of two perpendicular vectors is zero, since $\cos 90° = 0$. Thus $\mathbf{i} \cdot \mathbf{i} = \mathbf{j} \cdot \mathbf{j} = \mathbf{k} \cdot \mathbf{k} = 1$, whereas $\mathbf{i} \cdot \mathbf{j} = \mathbf{j} \cdot \mathbf{k} = \mathbf{k} \cdot \mathbf{i} = 0$.

An example of the use of a scalar product is given by the interaction of the magnetic moment μ of a nucleus with an applied field $\mathbf{B_0}$. The energy of this interaction is $E = -\mu \cdot \mathbf{B_0}$, which leads to the result $E = -\gamma \hbar m B_0$ given in eqn (5.6). Perhaps a more familiar example is the work W done in moving an object a distance ds against a force **F**: $W = -\mathbf{F} \cdot \mathrm{d}s$.

An alternative form of multiplication of two vectors is known as the *vector product* or *cross product*. The vector product of **p** and **q** is a vector of magnitude $pq \sin \theta$ and direction perpendicular to both **p** and **q**. The vector product is written **p** × **q**, and its direction is that in which a right-handed screw would move if turned from **p** to **q**. Thus **p** × **q** = −**q** × **p**, since the first vector product is equal to but opposite in direction to the second. Note that $\mathbf{i} \times \mathbf{i} = \mathbf{j} \times \mathbf{j} = \mathbf{k} \times \mathbf{k} = 0$ and that $\mathbf{i} \times \mathbf{j} = -\mathbf{j} \times \mathbf{i} = \mathbf{k}$.

An example of the vector product is given by the expression for the angular momentum **p** of an object about a point O. If **v** is the velocity of the object, m is its mass, and **r** is the position vector of the object relative to O, then the angular momentum **p** is given by

$$\mathbf{p} = \mathbf{r} \times m\mathbf{v}.$$

It can be seen from Fig. 5.16 that the direction of **p** is along the axis of rotation.

Another example of the vector product is given by the equation of motion of

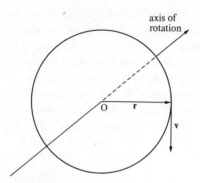

Fig. 5.16. Diagram illustrating that the direction of the angular momentum vector **p** is along the axis of rotation.

a magnetic moment μ in a field $\mathbf{B_0}$. The equation of motion is

$$\frac{\mathrm{d}\mu}{\mathrm{d}t} = \gamma\mu \times \mathbf{B_0}.$$

The motion represented by this equation is the precession of the magnetic moment about $\mathbf{B_0}$ with an angular frequency $\gamma\mathbf{B_0}$; thus the equation describes Larmor precession.

Appendix 5.2. Complex numbers

Consider the quadratic equation $x^2 - x - 1 = 0$. This can easily be solved to give $x = \{1 \pm \sqrt{(-3)}\}/2$, i.e. $x = \frac{1}{2} \pm i\sqrt{\frac{3}{2}}$ where $i = \sqrt{(-1)}$. The two solutions to this equation are complex numbers, comprising a real part ($\frac{1}{2}$) and an imaginary part ($\pm i\sqrt{\frac{3}{2}}$). The term imaginary implies that such numbers have no real physical meaning; nevertheless they play an extremely useful and elegant role in the analysis of waveforms. Complex numbers are therefore of considerable value in the theory of Fourier transform n.m.r., and it is primarily for this reason that this appendix is included.

Complex numbers are often represented by an Argand diagram in which the real part is represented by the x-coordinate, and the imaginary part by the y-coordinate, as shown in Fig. 5.17. The number $a + ib$ is thus totally specified by the vector $\overrightarrow{OP}$.

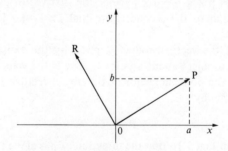

Fig. 5.17. An Argand diagram.

Let us consider multiplication of this number by i. The product is $ia + i^2 b$, which is equal to $ia - b$ since $i^2 = -1$. This is represented by the vector $\overrightarrow{OR}$. Note that the vector $\overrightarrow{OR}$ has the same magnitude as $\overrightarrow{OP}$, but has been rotated through $90°$. Multiplication by i corresponds to a rotation, or phase shift, of $90°$, and we shall see that this interpretation is not confined simply to the Argand diagram representation.

Let us consider the number $e^{i\phi}$, which can alternatively be written $\exp(i\phi)$. Now

$$e^x = 1 + x + \frac{x^2}{2} + \frac{x^3}{2 \cdot 3} + \frac{x^4}{2 \cdot 3 \cdot 4} + \ldots$$

(from which $e = 1 + 1 + 1/2 + 1/6 + 1/12 + \ldots = 2.71828$). $\exp(i\phi)$ can be expanded in a similar way to give

$$\exp(i\phi) = 1 + i\phi + i^2 \phi^2/2 + i^3 \phi^3/2 \times 3 + i^4 \phi^4/2 \times 3 \times 4 + \ldots$$
$$= 1 - \phi^2/2 + \phi^4/2 \times 3 \times 4 + \ldots$$
$$+ i(\phi - \phi^3/2 \times 3 + \ldots).$$

Now

and

$$\cos\phi = 1 - \phi^2/2 + \phi^4/2 \times 3 \times 4 + \ldots$$

$$\sin\phi = \phi - \phi^3/2 \times 3 + \ldots .$$

Therefore

$$\exp(i\phi) = \cos\phi + i\sin\phi.$$

Expressions of the form $A \exp(i\phi)$ are often used to describe waveforms for they contain information about amplitude (A) and about phase (ϕ). Moreover, such expressions are often far more amenable to mathematical analysis than cosine and sine terms.

It can easily be shown that

$$i \exp(i\phi) = \exp\{i(\phi + \pi/2)\}$$

and again it is interesting to note that multiplication by i is equivalent to a shift in phase by $90°$ or $\pi/2$ rad.

Appendix 5.3. The Fourier transform in n.m.r.

There is no unique definition of the Fourier transform, but a convenient formal definition is

$$F(\omega) = \int_{-\infty}^{\infty} f(t) \exp(i\omega t) \, dt .$$

The inverse transform is

$$f(t) = \frac{1}{2\pi} \int_{-\infty}^{\infty} F(\omega) \exp(-i\omega t) \, d\omega .$$

These expressions enable a time-dependent function $f(t)$ to be analysed in terms of its frequency components $F(\omega)$ and *vice versa*.

The response of the nuclear spins to a radiofrequency pulse can be expressed by

$$f(t) = A \cos \omega_0 t \exp(-t/T_2^*)$$

where ω_0 is the Larmor frequency and $\exp(-t/T_2^*)$ expresses the fact that the magnetization decays with a time constant T_2^*. The Fourier transform of this response is

$$F(\omega) = \int_0^\infty A \cos \omega_0 t \, \exp(-t/T_2^*) \exp(i\omega t) \mathrm{d}t$$

where the lower limit in the integral is zero because the time response is zero prior to the application of the pulse. On solving this integral we obtain a real component $F_{\text{real}}(\omega)$ given by

$$F_{\text{real}}(\omega) = \frac{A}{2} \frac{T_2^*}{1 + (T_2^*)^2 (\omega - \omega_0)^2}$$

and an imaginary component

$$F_{\text{imag}}(\omega) = \frac{iA}{2} \frac{\omega - \omega_0}{1 + (T_2^*)^2 (\omega - \omega_0)^2} .$$

The form of these two functions is shown in Fig. 5.11. The real part of the Fourier transform is precisely the Lorentzian lineshape given by eqn (1.7), and therefore in this example is equivalent to the absorption shape obtained in continuous-wave n.m.r. However, what about the second, imaginary term? We have seen in Appendix 5.2 that the imaginary term can be regarded as representing a component 90° out of phase with the real component. In n.m.r. the two terms correspond to the components of magnetization along two perpendicular axes of the rotating frame. Thus if the real component represents $M_{x'}$, the imaginary component represents $M_{y'}$. It turns out that these two terms are interdependent — the one specifies the other — and they have as much physical significance as each other. In the particular case we have considered the real part corresponds to the absorption mode and the imaginary part to what is called the dispersion mode (see below). However, if we had considered the transform of the decay $A \sin \omega_0 t \, \exp(-t/T_2^*)$, we would have found that the real part would now contain the dispersion mode while the imaginary part would contain a negative absorption signal (see Fig. 5.12). This corresponds to a 90° phase shift. In general, we transform a function of the form $A \cos(\omega_0 t + \phi)$ and obtain real and imaginary parts, each of which contains a mixture of the absorption and dispersion modes (see Fig. 5.13).

The dispersion mode is given its name for the following reason. In the optical region of the electromagnetic spectrum variation of the refractive index with wavelength or frequency has been known about for a long time and is called dispersion. Usually the refractive index increases with frequency and this is known as 'normal dispersion'. However, in the vicinity of an absorption line the reverse happens; the refractive index decreases with increasing frequency, a phenomenon known as 'anomalous dispersion'. In a simple case the absorption and dispersion are given by the two terms $F_{\text{real}}(\omega)$ and $F_{\text{imag}}(\omega)$ respectively that are given above; hence the terminology that is used.

6

The n.m.r. parameters and their measurement

In order to interpret any n.m.r. spectrum, it is essential to have some understanding of the various factors that influence the frequency, area, width, and shape of n.m.r. signals. In previous chapters, it was noted that the nature of n.m.r. signals is dependent on a wide range of physical and chemical effects, and the purpose of this chapter is to explain the basis of these effects, and to indicate how they can be quantified. We shall consider in turn (i) the chemical shift, (ii) spin–spin coupling, (iii) relaxation, (iv) concentration and signal intensity, (v) chemical exchange, and (vi) the nuclear Overhauser effect.

6.1. The chemical shift

Of the parameters that characterize n.m.r. signals, the chemical shift is undoubtedly the most important. The existence of the chemical shift enables us to use n.m.r. to distinguish not only between different molecules, but also between individual atoms within a molecule. When used in conjunction with intensity measurements and spin–spin coupling data, the chemical shifts of the spectral lines of a molecule provide a tremendous amount of information about its structure. N.m.r. is therefore an essential structural tool for organic chemists. It also provides an invaluable aid to structure determinations of biological molecules (see, for example, Campbell and Dobson 1979).

In § 1.4 a brief introduction was given to the concept of the chemical shift. In particular, it was noted that an applied field B_0 induces electronic currents in atoms and molecules, and that these produce an additional small field $B_{0\sigma}$ at the nucleus proportional to B_0. The total effective field at the nucleus can therefore be written

$$B_{eff} = B_0 - B_{0\sigma}$$
$$= B_0(1 - \sigma). \tag{6.1}$$

σ is called the shielding or screening constant because the effect of these electronic currents is to shield the nuclei from the effects of the applied field. σ is sensitive to the chemical environment of the nuclei, and therefore nuclei in different chemical environments experience different fields and hence produce signals at different frequencies. The separation of resonance frequencies from an arbitrarily chosen reference frequency is termed the chemical shift. The secondary shielding fields generated by the electrons are very small in comparison with the applied field, and so the absolute spread in frequency is also small. As a result, it can often be difficult to resolve the various resonances, particularly those of large molecules that produce complex spectra. Since the frequency separation of resonances is proportional to B_0, an increase in B_0 will tend to spread them out and hence improve the spectral resolution.

In this section we discuss some of the factors that contribute to chemical shifts and describe the practicalities and conventions involved in chemical shift measurements.

6.1.1. Conventions and terminology

(i) Chemical shifts are expressed in terms of the dimensionless unit of parts per million (p.p.m.), and as such are independent of the magnitude of the field B_0. The chemical shift δ is defined as

$$\delta = \frac{\nu_S - \nu_R}{\nu_R} \times 10^6 \tag{6.2}$$

where ν_S is the resonance frequency of the sample and ν_R is the frequency of an arbitrarily chosen reference. By using eqn (6.1), we find that

$$\delta = \frac{\sigma_R - \sigma_S}{1 - \sigma_R} \times 10^6$$

i.e.

$$\delta = (\sigma_R - \sigma_S) \times 10^6 \tag{6.3}$$

since $\sigma \ll 1$. Therefore the chemical shift is obtained simply from the difference between the shielding constants of the sample and of the reference nuclei. Equation (6.3) expresses the accepted convention that the chemical shift of a resonance is negative if the sample nucleus is more shielded than the reference. Unfortunately, many ^{31}P n.m.r. spectra have been published using the reverse convention, but throughout this book negative shifts correspond to increased shielding.

(ii) The terms 'upfield' and 'downfield' are frequently used to denote the direction of a chemical shift. These terms originated in the days of continuous-wave n.m.r. when spectra were obtained by using a fixed-frequency source and sweeping through the applied field B_0. Suppose that a nucleus that is weakly shielded resonates at a fixed frequency ν_0 in an applied field B_0. A nucleus that is more strongly shielded experiences a lower effective field, and therefore to bring it to resonance at the frequency ν_0 the applied field has to be increased. Its resonance is therefore said to be shifted upfield. If we now consider Fourier transform n.m.r., a fixed field is applied and resonances are separated according to their frequency. Again, the weakly shielded nucleus resonates at a frequency ν_0 in the field B_0. However, the more strongly shielded nucleus experiences a reduced field and must therefore resonate at a lower frequency. Thus an increase in shielding produces an *upfield* shift if the field is varied, but a shift to *low* frequency when frequency is the variable. Intuitively, one might have expected that, since frequency is proportional to field, upfield would have corresponded to high frequency; in fact, it corresponds to low frequency. Although the advent of Fourier transform n.m.r. has made the terminology of upfield and downfield shifts obsolete and confusing, these terms are still in common usage and therefore have to be tolerated for the present.

(iii) Spectra are plotted out according to the traditional rule of spectroscopy: increasing wavelength, or decreasing frequency, to the right. However, a number of ^{31}P n.m.r. spectra from the Oxford laboratory have been published with increasing frequency to the right; in this book these spectra are inverted in order to obey the convention of decreasing frequency to the right.

To summarize, if one nucleus is more shielded than another, its signal will be shifted to low frequency (or equivalently upfield); by convention it will have a more negative chemical shift, and will appear further towards the right-hand side of the spectrum.

6.1.2. Internal and external standards and the effects of magnetic susceptibility

The local field experienced by a nucleus is modified not only by the shielding effects of the local electrons but also by shielding produced by the surrounding medium. This 'bulk shielding' modifies the field by an amount B_{bs}

$$B_{bs} = S_f \chi B_0$$

where χ is the magnetic susceptibility[1] of the medium and S_f is a numerical factor that depends on the *shape* of the sample. The magnitude of χ is of the order of 10^{-5} for common materials, and therefore the bulk shielding cannot in principle be ignored.

Chemical shifts are measured by comparing an observed resonance frequency with that of a reference compound. When the reference is added directly to the sample, it is termed an *internal* reference. Under these conditions the reference and sample both experience the same bulk shielding and no corrections for susceptibility effects are necessary; observed chemical shift differences must be entirely due to local interactions. However, if it is not feasible to add the reference directly to the sample, it may be necessary to place within the sample tube a small capillary tube containing the reference compound. This is an *external* reference, and under these circumstances it may be necessary to correct the observed shift for the effects of susceptibility differences between reference and sample. Clearly, it is preferable, whenever possible, to use an internal reference, but of course a reference compound must possess a number of characteristics.

(i) Its spectrum should preferably consist of a single resonance that occurs well away from the sample resonances.

(ii) It should be chemically inert and its resonance frequency should be independent of the nature of the sample.

(iii) If possible, it should contain a large number of chemically equivalent nuclei in order that a low concentration should give an intense signal.

An accepted standard for ^1H n.m.r. is tetramethylsilane (TMS) which satisfies all of the above requirements. This compound is also an accepted standard for ^{13}C n.m.r. However, since TMS is miscible with organic solvents, but is only very

---

[1] Any substance acquires magnetization when placed in a magnetic field, as a result of interactions between its electrons and the field. χ is a measure of this magnetization, and it differs in sign according to whether the substance is diamagnetic or paramagnetic. In a diamagnetic substance, the induced magnetization opposes the applied field, and χ is negative; in contrast, for paramagnetic substances χ is positive. All substances show a diamagnetic effect (see § 6.1.3), whereas paramagnetism is associated with those molecules or ions (such as transition metal and lanthanide ions) that possess permanent electonic magnetic moments. If there is a paramagnetic contribution to χ it can far outweigh the diamagnetic contribution. (We know, of course, that a sample containing nuclei with magnetic moments develops a net magnetization when placed in a magnetic field. This 'nuclear paramagnetism' is totally analogous to the electronic paramagnetism that we are considering here, but is very much weaker because the nuclear moment is smaller than the electronic magnetic moment by a factor of the order of 1000.)

weakly soluble in water, it cannot be used as an internal reference for aqueous solutions. Therefore the water-soluble compound $(CH_3)_3$-Si-CH_2-CH_2-CH_2-SO_3Na (DSS) is often used for biological studies.

The traditional standard for ^{31}P n.m.r. is 85 percent phosphoric acid, but it should be stressed that this is a poor reference compound, particularly for studies of living systems. There are several reasons for this.

(i) Clearly, 85 percent phosphoric acid cannot be used as an internal reference.

(ii) The linewidth of phosphoric acid at 40 MHz is about 5 Hz (Glonek and van Wezer 1974), which is far greater than is desirable.

(iii) The bulk susceptibility of 85 percent phosphoric acid is very different from that of aqueous solutions. As a result, chemical shifts measured relative to phosphoric acid vary with sample shape and orientation, and as these properties differ more extensively in studies of living systems than in more traditional n.m.r. studies considerable variations in apparent chemical shifts can be expected. It is particularly important to appreciate that chemical shifts measured relative to phosphoric acid can differ by up to 1 p.p.m. according to whether a given cylindrical sample is studied using an electromagnet or a superconducting magnet; in the latter case the axis of the cylinder is usually (although not always) parallel to the applied field B_0, whereas for an electromagnet it is perpendicular to it. For example, the chemical shift of the phosphocreatine signal relative to phosphoric acid can vary between −2.3 p.p.m. and −3.2 p.p.m., depending *only* on the geometry of the sample. Unfortunately there is no commonly accepted alternative to phosphoric acid, although a number of suggestions have been made such as the tetrahydroxyphosphonium ion (Glonek and Van Wezer 1974).

The absence of a universally accepted standard does not mean that chemical shifts cannot be determined accurately; it simply means that great care has to be taken to make all the appropriate corrections, or that chemical shift measurements should be made relative to a convenient alternative reference and then be expressed on a commonly accepted chemical shift scale. For example, the ^{31}P signal of phosphocreatine in skeletal and cardiac muscle, and in brain, provides a convenient and acceptable internal reference. This is because phosphocreatine has a pK_a value of about 4.6, and its resonance frequency is therefore insensitive to pH changes within the normal physiological range. To convert chemical shifts measured relative to phosphocreatine to values relative to any other reference, it is simply necessary to perform an accurate comparison of the chemical shifts of the two compounds under the appropriate conditions.

6.1.3. *Contributions to chemical shifts*

What are the factors that contribute towards chemical shifts? We can distinguish between local and long-range contributions.

(i) *Local diamagnetic effect*

This effect, which we considered briefly in § 6.1, exists for all atoms and molecules, and arises because an applied field B_0 induces orbital motion in the electrons as shown in Fig. 6.1. This motion generates a secondary magnetic field at the nucleus that opposes B_0 and therefore 'shields' the nucleus from the applied field.

Fig. 6.1. The orbital motion of electrons, induced by the field B_0.

The magnitude of the local diamagnetic shielding is determined primarily by the local electron density. Therefore, if this is the dominant shielding mechanism, we should expect factors such as the electronegativity of adjacent atoms to be reflected in the magnitude of the observed shifts.

For protons, the local shielding is often dominated by the diamagnetic term, but for nuclei such as ^{13}C and ^{19}F, additional more complicated effects become important. These so-called paramagnetic effects (an explanation of which is far beyond the scope of this book) enhance the applied field B_0, in contrast to the diamagnetic contribution that opposes B_0.

(ii) *Long-range shielding*

The local effects considered above arise from shielding by electrons on the same atom as the nucleus in question. In this section we briefly discuss two examples of the long-range shielding caused by the induced motion of electrons associated with *other* atoms or groups of atoms. It can be shown that such long-range shielding occurs only if the motion of the neighbouring electrons is anisotropic, i.e. if the motion differs according to the orientation of the molecule in the applied field.

Ring-current shifts

In aromatic systems the delocalized π electrons give rise to a ring current if the applied field B_0 has a component perpendicular to the plane of the ring (see Fig. 6.2). This current generates a secondary field whose direction and magnitude at a

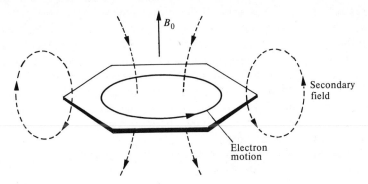

Fig. 6.2. Ring currents, and the secondary fields that they generate.

given nucleus is highly dependent on the orientation of the nucleus relative to the plane of the ring. For example an aromatic proton in the plane of the ring experiences an enhanced field, whereas a proton above or below the ring experiences a reduced field. Ring-current shifts can be large (of the order of 1 p.p.m.) and extend over considerable distances, typically up to about 6 Å.

Shifts from paramagnetic centres

Resonances of nuclei that are close to paramagnetic centres (such as transition metal or lanthanide ions) can exhibit quite remarkable shifts, which in general arise from one of two types of interaction.

(a) Contact shifts: these involve the transfer of unpaired electron density to a nucleus. This effect is usually transmitted via chemical bonds.

(b) Pseudo-contact or dipolar shifts: these result from a dipolar interaction between the electron spin and the nuclear spin. It is possible, at least in principle, to obtain information about the three-dimensional structure of a molecule by observing the various dipolar shifts caused by paramagnetic centres binding to it. The scope and difficulties of the method have been extensively reviewed (see for example, Dwek 1973, James 1975, Wuthrich 1976, Campbell and Dobson 1979), and are not discussed here.

It should be noted that paramagnetic centres broaden resonances in addition to shifting them. It turns out that some paramagnetic ions are particularly efficient at broadening resonances, whereas others produce significant shifts with relatively little broadening. Thus there are 'shift probes' such as Co^{2+}, Ni^{2+}, and most of the lanthanide ions, and 'broadening probes' such as Mn^{2+} and Gd^{3+}.

6.1.4. Chemical shift anisotropy

If a group within a molecule is anisotropic, the extent of shielding at a nucleus near or within the group will depend upon the orientation of the group relative to the applied field. We have noted this effect in the case of aromatic ring systems. There is no shielding if the plane of the ring is parallel to the applied field, but there are significant effects if the ring is perpendicular to the field. Thus the chemical shift of a nucleus near the ring will vary as the molecule containing the ring tumbles randomly in solution, and in fact will be averaged by the effects of motion to the value that is detected in the spectrum of the solution. This average, as we have seen, depends upon the precise position of the nucleus relative to the ring.

When the chemical shift of a nucleus depends on the orientation of the molecule within the applied field, the chemical shift is said to be anisotropic. When there is axial symmetry, the chemical shift anisotropy can be expressed in terms of the difference $\sigma_{||} - \sigma_{\perp}$, where $\sigma_{||}$ and $\sigma_{\perp}$ refer to the components of shielding parallel and perpendicular to the axis of symmetry of the grouping. Chemical shift anisotropy can make important contributions to the spectra of solids and highly immobilized systems such as membranes, because in such systems its effects are not averaged out by molecular motion.

Chemical shift anisotropy can also, as mentioned later (p. 115), provide a mechanism for spin—spin and spin—lattice relaxation.

6.2. Spin–spin coupling

It was mentioned in § 1.4 that resonances are often split into two or more components, as illustrated by the ^{31}P n.m.r. spectrum of ATP shown in Fig. 1.6. This splitting arises from an interaction between neighbouring nuclear spins which is transmitted via the electrons in the bonds joining the nuclei. An explanation for the effect can be given most simply by considering as an example two nuclei A and X, both of spin ½, in atoms that are joined by a covalent bond.

The possible spin states of these two nuclei are illustrated in Fig. 6.3. In general,

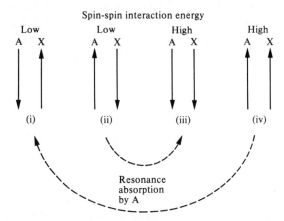

Fig. 6.3. This diagram illustrates the four possible spin states that exist for two coupled nuclei A and X, both of spin $\frac{1}{2}$. The $m = +\frac{1}{2}$ and $m = -\frac{1}{2}$ states are represented by upward-pointing and downward-pointing arrows respectively. The spin–spin interaction energy is higher in states (iii) and (iv) than in states (i) and (ii). Therefore less energy is absorbed in the transition from state (iv) to (i) than in the transition from (ii) to (iii).

it can be shown that the electron-mediated interaction between the two nuclei is more favourable (i.e. of lower energy) when the nuclear spins are antiparallel (as in (i) and (ii) in the figure) then when they are parallel, as in (iii) and (iv). Now the resonance absorption of energy by nucleus A involves a change in its spin from −½ to +½, with no change in the spin of X. Therefore it corresponds to a transition either from (ii) to (iii), or from (iv) to (i), as illustrated by the dotted lines in the figure. The amounts of energy absorbed in these two types of transition differ slightly from each other, because there is a net increase in the spin-spin interaction energy in one type, and a decrease in the other. Thus the resonance of nucleus A is split into two components, which are of equal intensities because the two types of transition are equally probable.

The separation of the two components is J_{AX} Hz, where J_{AX} is a constant known as the spin–spin coupling constant. J_{AX} is a characteristic of the molecule; it is independent of the magnitude of the applied field B_0, and this is why it is expressed in hertz rather than in parts per million. Since the interaction is mutual, the X resonance is also split into two components separated by J_{AX} Hz. Coupling constants have typical values in the region 1–200 Hz.

Spin–spin coupling can occur when two nuclei are bonded together, e.g. ^{13}C-^{1}H, or when they are separated by more than one bond, e.g. ^{1}H-C-C-^{1}H. The coupling is transmitted via the intervening chemical bonds but can be expected to decrease as the number of bonds increases. Often a superscript is used, e.g. $^{3}J_{AX}$, to indicate the number of bonds between the coupled nuclei. When there are more than two magnetic nuclei in a molecule, coupling occurs between each pair of nuclei and the splitting pattern can become very complicated.

Coupling patterns can be used empirically as a means of identifying chemical groupings within a sample. Often this identification relies upon double-resonance or spin-echo techniques (see below) which have been cleverly applied to proteins such as lysozyme (Campbell, Dobson, Williams, and Wright 1975) and the trypsin inhibitor (Wuthrich, Wagner, Richarz, and de Marco 1977). Alternatively, coupling patterns can be used as an aid to conformational analysis (see Feeney 1975; Bystrov 1976).

6.2.1. The analysis of n.m.r. spectra

The theoretical treatment of spin–spin coupling can only be considered in terms of quantum mechanics, and is outside the scope of this book. Here, therefore, we discuss the important rules governing spin–spin coupling and illustrate them by means of a few simple examples.

We begin by distinguishing between first-order and second-order spectra. First-order spectra are ones in which the frequency difference between the resonances of the nuclei involved in the spin–spin coupling is much greater than the magnitude of the coupling constant J. Second-order spectra, which are more difficult to interpret, arise when this condition is not fully fulfilled, i.e. when J is not small in comparison with the frequency difference.

There is an accepted nomenclature for spin systems whereby coupled nuclei that generate second-order spectra are assigned adjacent letters in the alphabet; otherwise they are assigned letters that are well separated. The letters conventionally used are A, B, and C, M and N, and X, Y and Z. Thus an AX system produces first-order spectra, while an AB system generates second-order spectra.

6.2.2 First-order spectra

The rules governing first-order spectra are as follows.

(i) The spin–spin coupling between the nuclei of an equivalent group, e.g. the three protons of the methyl group of acetaldehyde, produces no observable splitting. This is because of the existence of quantum-mechanical selection rules that prohibit the appropriate transitions.

(ii) A nucleus coupled to n equivalent nuclei of spin I gives rise to $2nI + 1$ lines with relative intensities given by the binomial distribution. Thus a nucleus coupled to a single nucleus of spin ½ would produce a doublet with relative intensities 1:1, coupling to two equivalent nuclei of spin ½ would produce a triplet with relative intensities 1:2:1, coupling to three equivalent nuclei of spin ½ would generate a quartet with relative intensities 1:3:3:1, etc.

(iii) If there is coupling to more than one group of interacting nuclei, a multiplet of multiplets is formed by straightforward extension of rule (ii).

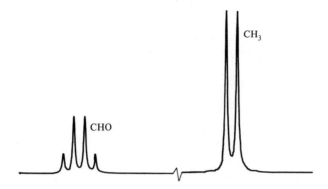

Fig. 6.4. ^{1}H n.m.r. spectrum of acetaldehyde (CH$_3$CHO). The signal from the CH$_3$ protons is split into a doublet through coupling to the CHO proton, and the CHO signal is a quartet because of coupling to the three CH$_3$ protons.

(iv) The spacing within each multiplet due to a given interaction is equal to the coupling constant J for that interaction.

The ^{1}H spectrum of acetaldehyde shown in Fig. 6.4 and the ^{31}P spectrum of ATP that was shown in Fig. 1.6 provide simple examples of first-order spectra. In the ATP spectrum the signal from the γ-phosphate is split into two by interaction with the ^{31}P nucleus of the β group. The signal from the β- phosphate is split into a doublet of doublets by the α and γ ^{31}P nuclei. Since the coupling to both of these nuclei is the same, this results in the appearance of a triplet of lines of relative intensities 1:2:1. The signal from the α-phosphate is split into a doublet by the ^{31}P nucleus of the β group, and there is a smaller interaction with the CH$_2$ protons of the adenine moiety which would, if the resolution were sufficiently good, produce a subsidiary splitting of the two doublet lines.

6.2.3. Second-order spectra

Typical second-order spectra for two coupled nuclei A and B, both of spin ½, are shown in Fig. 6.5. This figure illustrates what happens to the splittings and relative intensities as the frequency separation between the A and B resonances gradually increases while the spin–spin coupling remains constant. This is the sort of effect that one would expect on gradually increasing the magnetic field B_0 at which the experiment is performed.

The frequencies of the lines relative to the centre of the pattern and the intensities of the lines are given in Table 6.1. Note that the splitting between the pair 1 and 2 and between the pair 3 and 4 is always equal to J.

6.2.4. Spin decoupling

The ^{13}C signal from, say, the CH$_3$ groups of acetone is split into four components as a result of spin–spin coupling with the methyl protons. It can be shown, both theoretically and experimentally, that if irradiation is applied at the resonance frequency of the methyl protons, the ^{13}C quartet is collapsed into a single line. This

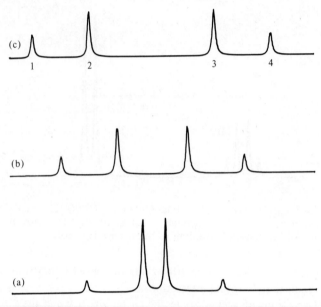

Fig. 6.5. Second-order spectra for two coupled nuclei A and B, both of spin $\frac{1}{2}$. The frequency separation $\nu_A - \nu_B$ is equal to (a) J; (b) $2J$; (c) $3J$. If the frequency separation were negligible in comparison with J, we would obtain a single central line; if it were very large in comparison with J, we would obtain two doublets, with all four lines of equal intensity (see rules (i) and (ii) in § 6.2.2). ν_A and ν_B are the frequencies A and B would generate if there were no spin–spin coupling.

Table 6.1

Frequencies and intensities of the lines in an AB spectrum.

Line number	Frequency	Intensity
1	$\frac{1}{2}J + C$	$1 - \sin 2\theta$
2	$-\frac{1}{2}J + C$	$1 + \sin 2\theta$
3	$\frac{1}{2}J - C$	$1 + \sin 2\theta$
4	$-\frac{1}{2}J - C$	$1 - \sin 2\theta$

C is equal to $\frac{1}{2}\sqrt{[J^2 + (\nu_A - \nu_B)^2]}$, and θ is given by $\tan 2\theta = J/(\nu_A - \nu_B)$. The pattern is centred about the frequency $(\nu_A + \nu_B)/2$.

process of spin decoupling occurs provided that the amplitude of the irradiating field (generally termed B_2) is such that $\gamma B_2 \gg nJ$, where J is the coupling constant and n is the number of lines in the multiplet.

Spin decoupling is very widely used and is of tremendous value as a means of both assigning and simplifying spectra. In ^{13}C n.m.r. experiments it is common practice to use 'broad-band proton decoupling', i.e. to apply the second radiofrequency field B_2 in such a way that the whole of the ^1H spectral range is irradiated. This causes a collapse of many of the ^{13}C multiplets into singlets, which provides extensive simplification of the spectra. However, this is by no means the only reason for using double irradiation, for the B_2 field can also saturate the ^1H resonances and thereby lead to large enhancements of the ^{13}C signal intensities through the nuclear Overhauser effect. This effect is discussed in § 6.6.

6.3. Relaxation

When a sample is placed in a magnetic field it becomes magnetized, as was shown in Chapter 5. At equilibrium the component of magnetization along the field (the z axis) is equal to M_0, whereas the net magnetization M_{xy} perpendicular to the field is zero. Following any perturbation of this magnetization, e.g. following the application of a 90° pulse, processes take place whereby M_z and M_{xy} return to their equilibrium values of M_0 and zero respectively. It is essential to have some understanding of these so-called relaxation processes because of their critical role in the theory and practice of n.m.r. spectroscopy.

6.3.1. *Spin–lattice relaxation*

The return of M_z to its equilibrium value is termed spin–lattice relaxation and is characterized by a time constant T_1 known as the spin–lattice or longitudinal relaxation time. The term spin–lattice is used because the processes involve an exchange of energy between the nuclear spins and their molecular framework, which is referred to as the lattice regardless of the physical state of the system.

The return of the nuclear spins to thermal equilibrium, i.e. to equilibrium with the lattice, is often, although not always, exponential. If it is exponential, we can write a simple equation[1] describing the relaxation

$$\frac{dM_z}{dt} = \frac{M_0 - M_z}{T_1} \qquad (6.4)$$

This equation expresses the return of the magnetization M_z to its equilibrium value M_0 with a time constant T_1.

Let us now consider the nature of the processes that are responsible for relaxation. We have seen that the relative populations of the spin states can be altered in a well-defined way by application of a resonant B_1 field applied in the xy-plane. In a similar manner, *any* fluctuating magnetic field which has a component in the xy-plane that oscillates at the resonant frequency will induce transitions between the spin states of the nuclei. If these fluctuating fields are associated with the lattice, there will be an exchange of energy until the nuclear spins are in thermal equilibrium with the lattice. Thermal equilibrium is characterised by the temperature T of the lattice, and the relative populations of the nuclear energy states at equilibrium will be those characteristic of this temperature, i.e.

$$n^-/n^+ = \exp(-\Delta E/kT). \qquad (6.5)$$

What processes generate these fluctuating magnetic fields? As an example, consider two nuclei A and X, both of spin ½, within a molecule. Since X has a magnetic

---

[1] This equation and eqn (6.8) are simplified forms of the well-known Bloch equations (see Bloch 1946).

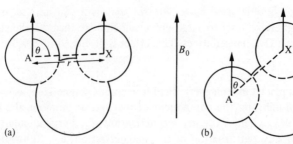

Fig. 6.6. Two nuclei A and X within a molecule both have their spins parallel to B_0. The field at A due to the spin of X is dependent on the angle θ defined as shown in the diagram. As the molecule rotates from (a) to (b), the angle θ changes, and hence the interaction between A and X also changes. r is the distance between A and X.

dipole moment, it produces a field at A whose component B_{xy} in the xy-plane is $\frac{3}{2} \sin\theta \cos\theta \; \gamma/\hbar r^3$ (see Fig. 6.6, where θ and r are defined). If the molecule is tumbling around in solution, the relative orientations of A and X change randomly. As a result, the angle θ fluctuates, and hence the field B_{xy} also fluctuates. If there are any components of molecular motion that happen to fluctuate at the resonance frequency, then these can cause relaxation of nucleus A. The two nuclei do not have to be in the same molecule; for example, diffusion of molecules can modulate fields by changing both θ and r, and can therefore lead to relaxation. Relaxation generated by this interaction between neighbouring nuclear magnetic dipole moments is termed dipole–dipole relaxation.

The frequency distribution of the motion of a randomly tumbling molecule can be expressed in terms of the *spectral density* $J(\omega)$, and under many circumstances is given by

$$J(\omega) = \frac{\tau_c}{1 + \omega^2 \tau_c^2} \tag{6.6}$$

where τ_c is known as the correlation time. τ_c expresses the characteristic timescale of the molecular motion. For example, at room temperature water molecules in solution have a correlation time of about 3×10^{-12} s, ATP might be expected to have a correlation time of about 10^{-10} s, and an enzyme of molecular weight 20 000 might have a correlation time of about 10^{-8} s.

We might expect that the relaxation rate $1/T_1$ would be dependent upon the magnitude of the fluctuating fields and upon the spectral density at the resonance frequency ω_0, and indeed it can be shown that

$$\frac{1}{T_1} \alpha \, B_{xy}^2 \, \frac{\tau_c}{1 + \omega_0^2 \tau_c^2} \tag{6.7}$$

The variation of this expression with τ_c is such that, as expected, this relaxation mechanism is most efficient when $\tau_c = 1/\omega_0$, i.e. when the characteristic frequency $(1/\tau_c)$ of the molecular motion is equal to the resonance frequency. However, it should be pointed out that there is an additional contribution to dipole-dipole relaxation which makes the full expression for T_1 rather more complicated (see § 6.3.3).

6.3.2. Spin–spin relaxation

The return of M_{xy} to its equilibrium value is termed spin–spin relaxation and is characterized by a time constant T_2 known as the spin–spin or transverse relaxation time. The term spin–spin is used because the relaxation processes involve interactions between neighbouring nuclear spins without any exchange of energy with the lattice. The equilibrium value of M_{xy} is zero as there is no preferred orientation within the xy-plane, and therefore the equation describing the return to equilibrium may be written

$$\frac{dM_{xy}}{dt} = -\frac{M_{xy}}{T_2} \qquad (6.8)$$

This equation expressed the exponential decay of M_{xy} to zero with a time constant T_2.

   Consider the magnetization M_{xy} generated by a radiofrequency pulse. If there is a single resonance frequency ω_0, the magnetization precesses coherently about the z axis with frequency ω_0; in the rotating frame of reference the magnetization remains static and of constant magnitude. However, if there is a spread $\Delta\omega_0$ of frequencies, different nuclei will precess at slightly different frequencies; in the rotating frame the nuclear magnets will 'fan out', as shown in Fig. 6.7, the net

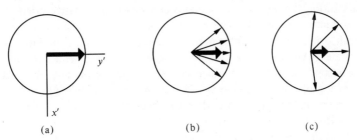

(a)   (b)   (c)

Fig. 6.7. The magnetization in the $x'y'$ plane of the rotating frame. (a) represents the situation immediately after the application of a 90° pulse; the net magnetization is represented by the heavy arrow. The individual nuclear magnets gradually fan out, as shown by the light arrows in (b), and then (c). Therefore, the *net* magnetization in the $x'y'$ plane gradually declines, the time constant being given by T_2. Additional effects such as B_0 inhomogeneity cause the net magnetization in this plane to decay even more rapidly.

effect being that M_{xy} decays. The greater the frequency spread, the more rapidly M_{xy} decays. If T_2 represents the time constant of the decay of M_{xy} that results from spin–spin relaxation, then we find that

$$\frac{1}{T_2} = \frac{\Delta\omega_0}{2} = \pi\Delta\nu_{1/2} \qquad (6.9)$$

where $\Delta\nu_{1/2}$ is the corresponding resonance linewidth. Spin–spin relaxation therefore involves processes that cause an inherent broadening of the resonance linewidths. We now consider the nature of these processes.

Firstly, we note that, as a result of spin–lattice relaxation processes, the nuclear spins have a finite lifetime in a given energy state. There is therefore an inherent uncertainty in the resonance frequency, $\Delta\omega_0 \approx 1/T_1$, and therefore an inherent 'lifetime' broadening of the resonances by this amount. Thus all processes that contribute to spin–lattice relaxation also affect T_2, and it turns out that T_2 cannot be longer than T_1.

In biological n.m.r. the lifetime broadening is often much less important than the effects we now consider. When there are two neighbouring nuclei A and X, both of spin ½, we have seen that the fluctuating fields in the xy-plane produced by X at A cause spin–lattice relaxation of the A nucleus. Now consider the component of field B_z in the z direction produced at A by X. The magnitude of B_z is ½$(3\cos^2\theta - 1)\gamma/\hbar r^3$ (see Fig. 6.6., where θ and r are defined). Thus B_z depends upon the relative orientations of A and X. If A and X are stationary, for example if they are in a powdered solid, then A will resonate at a frequency determined by its fixed orientation in space relative to X. In other molecules within the solid A and X have different relative orientations, and therefore A will resonate at different frequencies. The net result is that the solid produces a very broad resonance, which can be regarded as a summation of a large number of resonances spread over a range of frequencies. The range of frequencies $\Delta\nu$ is given by $\Delta B_z \approx h\Delta\nu$ where ΔB_z is the range of B_z fields experienced by A. $\Delta\nu$ might be typically about 5 kHz.

In a solution the motion of the molecules causes the relative orientations of A and X to fluctuate randomly. This motion tends to average the interaction of A and X over all orientations. If there were complete averaging of the interaction, then each nucleus A would experience the same time-averaged field from X and therefore the resonance from A would become a single narrow line. We might expect that averaging of the interaction would become more complete as the speed of the motion increases and therefore that the linewidths would decrease as motion increases. This 'motional narrowing' indeed occurs, and it is found that the linewidth in solution is proportional to the correlation time τ_c for the motion under consideration. The linewidth is also proportional to the square of the fluctuating field B_z. Thus

$$1/T_2 \, \alpha \, B_z^2 \tau_c. \tag{6.10}$$

A fairly simple derivation of this relationship is given by Slichter (1963).

Note that the expression $J(\omega) = \tau_c/(1 + \omega^2\tau_c^2)$ simplifies to $J(\omega) = \tau_c$ for $\omega \approx 0$. Thus eqn (6.10) for T_2 is exactly analogous to eqn (6.7) for T_1, except for the following.

(i) Equation (6.10) involves the z component of the fluctuating field rather than the component in the xy-plane.

(ii) Equation (6.10) involves the spectral density at $\omega \approx 0$, rather than at $\omega = \omega_0$. Thus the components of motion at low frequencies contribute to T_2 but not to T_1. However, it should be recalled that T_2 also contains a contribution from those components of motion at $\omega = \omega_0$ that also contribute to T_1. This is the lifetime broadening effect that was considered above.

6.3.3. T_1, T_2, and molecular mobility

From the above discussion, we have seen that fluctuating magnetic fields associated with molecular motion are responsible for both spin–spin and spin–lattice relaxation, but that the timescale of these fluctuations determines their relative contributions to the two types of relaxation. In particular, slow motions contribute only to spin–spin relaxation, whereas components of motion at the resonance frequency contribute to both spin–spin and spin–lattice relaxation. However, the situation is rather more complicated than this, because it is possible for the two nuclei A and X to undergo simultaneous transitions. Both nuclei can undergo transitions in the *same* direction if there is a component of molecular motion at frequency equal to the *sum* of the Larmor frequencies of the two nuclei. Such a process will contribute to spin–lattice relaxation, and hence by lifetime broadening to spin–spin relaxation also. If A and X are like nuclei, they can also undergo spin–spin exchange in which a downward transition of one nucleus is accompanied by a simultaneous upward transition of the other without any exchange of energy with the lattice. Such a process does not affect T_1, but it does affect the lifetimes of the nuclear spin states and therefore contributes to T_2. In fact, this contribution to T_2 has a similar form to eqn (6.10).

The calculation of T_1 and T_2 is not at all straightforward, but we can illustrate the above discussion by simply quoting the T_1 and T_2 values for the protons of the

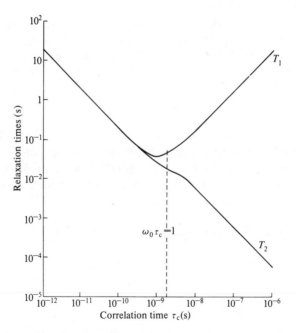

Fig. 6.8. Diagram showing how the relaxation times of the protons in H_2O vary with correlation time τ_c at a resonance frequency of 100 Mz. The graphs are based on eqns (6.11) and (6.12).

water molecule. The values are calculated by considering the dipole–dipole inter-action between the two protons of each individual molecule and ignoring the effects of protons on other molecules. The results are

$$\frac{1}{T_1} = \frac{3}{10} \frac{\gamma^4 \hbar^2}{r^6} \left(\frac{\tau_c}{1 + \omega_0^2 \tau_c^2} + \frac{4\tau_c}{1 + 4\omega_0^2 \tau_c^2} \right) \tag{6.11}$$

$$\frac{1}{T_2} = \frac{3}{20} \frac{\gamma^4 \hbar^2}{r^6} \left(3\tau_c + \frac{5\tau_c}{1 + \omega_0^2 \tau_c^2} + \frac{2\tau_c}{1 + 4\omega_0^2 \tau_c^2} \right) \tag{6.12}$$

The various terms in these expressions correspond to the different types of inter-action that were discussed above. The variation of T_1 and T_2 with correlation time, as deduced from eqns. (6.11) and (6.12), is given in Fig. 6.8.

When $\omega_0 \tau_c \ll 1$, it can be seen from Fig. 6.8 that $T_1 = T_2$, and the absolute values of T_1 and T_2 are estimated to be about 7 s at room temperature. This is con-sistent with the measured values of 3.6 s for both T_1 and T_2, as other mechanisms also contribute to the observed relaxation.

When $\omega_0 \tau_c \gg 1$, T_2 becomes much shorter than T_1. A small ratio of $T_2 : T_1$ is common in biological n.m.r. This is largely because the correlation times of bio-logical molecules are long as a result of their large size. In addition, there are other mechanisms that tend to affect linewidths more strongly than T_1; these include scalar coupling (see § 6.3.4) and chemical exchange (see § 6.5).

6.3.4. Relaxation mechanisms

In the preceding sections we have considered only one type of relaxation mech-anism – that generated by the interaction between neighbouring nuclear magnetic dipole moments. In fact, there are many different mechanisms that can contribute to the observed relaxation. If two or more independent relaxation mechanisms exist, the net effect is obtained from expressions of the following type:

$$(1/T_1)_{\text{obs}} = 1/T_{1A} + 1/T_{1B} + 1/T_{1C} + \ldots \tag{6.13}$$

where $T_{1\,\text{obs}}$ is the observed spin–lattice relaxation time, and T_{1A}, T_{1B}, T_{1C}, etc. are the relaxation times that characterize the individual relaxation mechanisms. A similar expression is obtained for the spin–spin relaxation time. Since the linewidth $\Delta\nu_{1/2}$ is proportional to $1/T_2$, the expression for T_2 is equivalent to the following:

$$\Delta\nu_{1/2}^{\text{obs}} = \Delta\nu_{1/2}^{A} + \Delta\nu_{1/2}^{B} + \Delta\nu_{1/2}^{C} + \ldots \tag{6.14}$$

where $\Delta\nu_{1/2}^{A}$, $\Delta\nu_{1/2}^{B}$, $\Delta\nu_{1/2}^{C}$, etc. are the characteristic linewidths generated by each relaxation mechanism.

In this section we shall discuss the most important relaxation mechanisms and indicate the extent to which they might be expected to contribute to the observed relaxation.

The dipole–dipole interaction

This interaction is the most important source of relaxation for many nuclei of spin ½ and can be expected to provide the dominant relaxation mechanism for the majority of ^{1}H, ^{13}C, and ^{31}P nuclei in biological molecules. Exceptions are most likely to be nuclei which are far removed from any other nuclear spins. This is because the dipole–dipole interaction has a strong $(1/r^6)$ distance dependence. It should be noted that ^{1}H nuclei have much larger magnetic dipole moments than ^{13}C and ^{31}P nuclei, and therefore the dipole–dipole relaxation tends to be more efficient for protons than for ^{13}C and ^{31}P nuclei.

Chemical shift anisotropy

Chemical shift anisotropy, as discussed in § 6.1.4, describes situations in which the chemical shift of a nucleus depends on the orientation of its molecular environment in the applied field B_0. As the molecule tumbles randomly in solution, the chemical shift and hence the local field will fluctuate, and this local field fluctuation provides a relaxation mechanism. When there is axial symmetry the contributions of this mechanism to T_1 and T_2 are given by

$$\frac{1}{T_1} = \frac{2}{15} \gamma^2 B_0^2 (\sigma_{||} - \sigma_\perp)^2 \left(\frac{2\tau_c}{1 + \omega_0^2 \tau_c^2} \right) \tag{6.15}$$

$$\frac{1}{T_2} = \frac{1}{45} \gamma^2 B_0^2 (\sigma_{||} - \sigma_\perp)^2 \left(4\tau_c + \frac{3\tau_c}{1 + \omega_0^2 \tau_c^2} \right) \tag{6.16}$$

The noteworthy feature of this mechanism is that its effects are proportional to B_0^2. As a result, we can expect the mechanism to become increasingly significant as the field B_0 increases.

Values for $\sigma_{||} - \sigma_\perp$ often have the same order of magnitude as the chemical shift ranges that characterize the nuclei. For example, $\sigma_{||} - \sigma_\perp$ is typically a few p.p.m. for ^{1}H, and 100–300 p.p.m. for ^{19}F, ^{13}C, and ^{31}P. The linewidths of some ^{31}P signals from living systems increase with increasing field strength, and it seems reasonable to interpret this effect in terms of a significant contribution from chemical shift anisotropy (see Gadian *et al.* 1979). In addition, it has been shown that for non-protonated unsaturated carbon atoms this mechanism can be important for ^{13}C nuclei at high field strengths (see Egan, Shindo, and Cohen 1977).

An unfortunate consequence of chemical shift anisotropy is that it can significantly reduce the gain in resolution that might be expected on increasing the magnetic field strength. If this mechanism is dominant, the linewidth will increase according to B_0^2, whereas the frequency range of the spectrum increases according to B_0 and the resolution will actually decrease on increasing the magnetic field. The best resolution is obtained at a field such that chemical shift anisotropy accounts for half the linewidth and is not necessarily obtained at the highest field that might be available.

Spin rotation

When a molecule rotates the electrons participating in this rotation produce a local magnetic field at the nuclei, the magnitude of which is dependent on the rotational angular momentum of the molecule. Brownian motion causes this angular momentum to fluctuate, and the local field fluctuations that are thereby generated can cause relaxation. However, this mechanism of spin rotation is likely to be more significant for small than for large molecules and is unlikely to produce significant relaxation in large biological molecules.

Scalar coupling

The spin–spin coupling interaction (see § 6.2) is alternatively known as the scalar interaction because its strength is independent of the orientation of the molecules within the applied field. Modulation of the magnitude of this interaction can generate relaxation. The fluctuations are usually *slow* on the timescale we must adopt when considering relaxation, and so the effect contributes far more significantly to spin–spin than to spin–lattice relaxation. However, relaxation via scalar coupling is unlikely to be of widespread importance for biological molecules, except in the case of paramagnetic systems where the coupling is between a nucleus and an electron. Such coupling is then referred to as a contact interaction.

Electric quadrupole relaxation

Nuclei with spin ½ have a spherical charge distribution. This is not true for nuclei of spin greater than ½, and such nuclei possess an electric quadrupole moment. The electric quadrupole moments interact with local electric field gradients. Relaxation is generated by fluctuations in the strength of this interaction resulting from molecular motion. This can be a very efficient relaxation mechanism and it accounts for the broadness of many of the resonances from quadrupolar nuclei.

Relaxation in paramagnetic systems

Paramagnetic centres generate very powerful relaxation because the magnetic moment of the electron is about 1000 times greater than the nuclear moments. This leads to an enhancement in relaxation by a factor of about 10^6, and so even trace amounts of paramagnetic ions can have a profound effect on relaxation times (see § 2.3).

Because of the strong distance dependence of the relaxation, paramagnetic ions can be used to probe the spatial distribution of nuclei within a molecule. Note that paramagnetic ions can also generate shifts in the resonance frequencies of nuclei (see p. 104). As a result, conformational studies are best performed by combining information from 'shift' and 'relaxation' probes. An excellent treatment of the effects of paramagnetic ions on nuclear resonances can be found in the book by Dwek (1973).

Molecular oxygen is paramagnetic, and it is important to be aware of the possible contribution that dissolved oxygen might make to the observed relaxation times. Chiarotti, Cristiani, and Giuletto (1955) showed that earlier estimates of 2.3 s for the ^1H spin–lattice relaxation time of H_2O contained a contribution from dissolved oxygen. They obtained a value of 3.6 s on 'degassing' the sample. However, in general the effects of dissolved oxygen on the relaxation times of biological molecules are very small and degassing of samples is rarely necessary.

6.3.5. Why measure T_1 and T_2?

Before discussing the methods of measuring T_1 and T_2, it is appropriate to discuss briefly why one should wish to make these relatively laborious and time-consuming measurements. There are many reasons, some of which are given below.

(i) In order to select the most suitable radiofrequency pulse interval it is essential to know the T_1 values of the various resonances, or equivalently their saturation factors (see § 7.2.4).

(ii) Reaction rate constants can be determined from saturation transfer experiments provided that the T_1 values of the resonances are known (see Appendix 6.1).

(iii) In principle, T_1 and T_2 measurements can provide information about molecular structure and mobility. In practice, the most detailed information is available from experiments involving the use of paramagnetic probes. However, it should be stressed that in general interpretation of the observed relaxation times can be extremely difficult: for living systems only the most qualitative of conclusions can be obtained about mobility and little if any information has emerged about structure.

(iv) Observed linewidths and lineshapes can be compared with the contribution to linewidths resulting from spin–spin relaxation. If there is a discrepancy between the two, there must be additional contributions to the observed linewidths. These may be from magnetic field inhomogeneities, or alternatively there may be more interesting explanations. For example, the width of the inorganic phosphate resonance in experiments on rat muscle has been interpreted in terms of a pH distribution with the tissue (Seeley *et al.* 1976).

(v) T_2 measurements using different intervals between $180°$ pulses can provide information about diffusion rates, and sequences that are used for T_2 measurements also provide a method of simplifying spectra and enhancing resolution (see § 6.3.7.).

(vi) T_1 provides a valuable contrast parameter in imaging studies (see Chapter 4).

For these and numerous other reasons measurements of relaxation times can be of great value, as indicated by the large number of papers on the subject that appear every year.

6.3.6 *Measurement of* T_1

There are several different radiofrequency pulse sequences that can be used for measuring spin–lattice relaxation times. A popular method, known as inversion recovery, is illustrated in Fig. 6.9. The application of a $180°$ pulse inverts the magnetization so that it is directed along the negative z axis. The magnetization relaxes back towards its equilibrium value M_0 with a time constant T_1, and a $90°$ pulse applied after a time τ samples the value to which M_z has relaxed after this time. Following signal detection, the system is allowed to return to equilibrium by waiting for a time t_D equal to at least $4T_1$ and the sequence is then repeated until a sufficient signal-to-noise ratio is obtained. The accumulated signal is Fourier transformed in the usual way, and the whole procedure is performed for a variety of different τ values. T_1 can then be obtained from a plot of $\ln(M_\infty - M_\tau)$ against τ which should be a straight line of gradient $1/T_1$.

A feature of the inversion recovery method is that the slope of the plot is little affected by any inaccuracies in the $180°$ and $90°$ pulses that may result from incorrect setting of the pulse widths or, more commonly, from inhomogeneities in

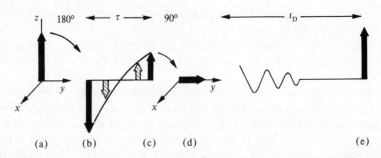

Fig. 6.9. Measurements of T_1 using the inversion-recovery method. (a) Initially, the net magnetization is equal to M_0 and is directed along the z-axis. (b) Following a $180°$ pulse which inverts the magnetization, $M_z = -M_0$. M_z then recovers towards its initial value M_0 with a time constant T_1. (c) after time τ the magnetization has reached a certain value as shown. (d) A $90°$ pulse tilts this magnetization into the xy-plane. The size of the resulting signal is determined by the magnetization at time τ. (e) After time t_D ($\geqslant 4T_1$) M_z has relaxed to its initial value of M_0, and the cycle can then be repeated.

the B_1 field. However, it is a very time-consuming method which is a particular drawback when studying systems that can only be kept in a steady state for a limited period of time.

The inversion recovery method can be written in the notation $(180° - \tau - 90° - \text{AT} - t_D)_n$, where AT represents the time for which the free-induction decay is acquired and n is the number of accumulations that are made for each value of τ. An alternative method, known as saturation recovery, can be written $(90° - \text{HS} - \tau - 90° - \text{AT} - \text{HS})_n$ where HS represents a homogeneity-spoiling pulse or 'homo-spoil'; this is a pulse that can be applied to one of the B_0 field homogeneity coils. Its effect is to destroy the B_0 homogeneity for a brief period and hence to destroy all magnetization in the xy-plane. In the saturation recovery method a $90°$ pulse tilts the magnetization into the xy-plane and then a homospoil pulse is applied. The net effect should be that there is no net magnetization along any direction so that the system is well and truly saturated. After a time τ a second $90°$ pulse is applied and the signal is monitored. Another homospoil pulse is applied, for the same reason as previously, and the whole process is repeated. Again, a suitable semi-logarithmic plot enables T_1 to be measured. The main advantage of the method is that T_1 values can usually be measured much more rapidly than with the inversion recovery method; this is because the long delay time $t_D \geqslant 4T_1$ is not required. In addition, the saturation recovery method is less susceptible than inversion recovery to problems that may arise from limitations in the power of the radiofrequency pulses.

If speed of measurement is to be the major criterion for choice of method, then the progressive saturation technique, which can be written $(90° - \text{AT} - t_D)_n$, should be considered. Essentially, this just involves applying a train of $90°$ pulses and collecting signal in the conventional manner. After a sufficient signal-to-noise ratio has been acquired at a given time interval, the interval is changed and the process is repeated. Again, a semi-logarithmic plot gives T_1. There are two main problems with the method, both of which can often be overcome. Firstly, the measured T_1 value is sensitive to the accuracy of the $90°$ pulse; incorrect values will be obtained if the pulse is mis-set or if there is significant inhomogeneity in the B_1 field. However, the accuracy of $90°$ pulses can be greatly enhanced by using 'composite pulses' of the type described by Freeman, Kempsall, and Levitt (1980). Secondly, the time interval between consecutive pulses has a minimum value equal to the acquisition time. This creates difficulties if T_2 is not much shorter than T_1 because the length of the acquisition time severely limits the range of pulse intervals that can be used. Under these circumstances the acquisition time could be shortened, but this would adversely affect the spectral resolution. In addition, a homospoil pulse would have to be applied to destroy magnetization in the xy-plane prior to application of each radiofrequency pulse. However, for biological systems the signals are relatively broad and the acquisition time relatively short, and so this particular problem rarely arises. Finally, it is essential to ensure that the magnetization reaches a steady state prior to the start of each accumulation, and for this reason signal should not be acquired from the first of a train of pulses.

There is a vast literature on the measurement of spin–lattice relaxation times, and the reader is referred to the article by Levy and Peat (1975) for a detailed account of the relative merits of the various methods.

6.3.7. Measurement of T_2

If no factors other than relaxation contribute towards resonance linewidths, then the spin–spin relaxation time T_2 can be obtained from the relationship $1/T_2 = \pi \Delta \nu_{1/2}$, where $\Delta \nu_{1/2}$ is the width at half-height. However, effects such as B_0 inhomogeneity often make additional contributions to the observed linewidths and we must resort to cleverer methods for estimating T_2. The 'spin-echo' method first proposed by Hahn (1950) employs a $90° - \tau - 180°$ pulse sequence and is illustrated in Fig. 6.10.

The $90°$ pulse tilts the magnetization into the xy-plane, or alternatively into the $x'y'$-plane of the rotating frame of reference. If all the spins were to precess at exactly the same frequency, then in the rotating frame of reference the magnetization would remain static along the y'-axis. However, as a result of B_0 inhomogeneity, nuclei in different regions of the sample experience slightly different fields and therefore precess at slightly different frequencies. In the rotating frame of reference this will result in a 'fanning out' of the magnetization as shown in Fig. 6.10(c). There is a gradual loss of coherence, and therefore of signal, as observed in a free-induction decay. Now if a $180°$ pulse is applied at a time τ, the various components of the magnetization will be inverted as shown, and provided that each component continues to precess at the same frequency (as it will if it stays in the same field strength) all the components of magnetization will arrive at the $-y'$-axis at the same time. This results in the formation of an 'echo' signal at a time τ after the $180°$ pulse. Since the echo is formed along the negative y'-axis, the signal will be of opposite phase to that observed after the $90°$ pulse. The echo should have the same amplitude as the initial signal provided that no relaxation occurs during the

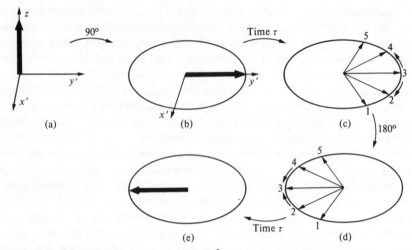

Fig. 6.10. The spin-echo experiment. A $90°$ pulse tilts the magnetization on to the y'-axis, as shown in (b). (c), during the time τ the dephasing takes place; (d), a $180°$ pulse about the x'-axis tilts the components of magnetization as shown; (e) *if* the various components continue to precess at the same speed, they come together to form an echo along the $-y'$-axis after an additional time τ.

time 2τ and provided that the effects of diffusion of molecules during this time are small (see below). However, if there is any dephasing or 'fanning out' of the magnetization during the time 2τ as a result of relaxation processes, this cannot be refocused by the $180°$ pulse because of the random nature of these processes. Therefore the amplitude of the echo will be diminished by a factor $\exp(-2\tau/T_2)$. Thus the amplitude of the echo enables T_2 to be evaluated.

This relatively simple procedure can be extended by using a train of $180°$ pulses applied at intervals of 2τ. Each $180°$ pulse generates an echo as described above, and if signals were detected midway between each pulse a train of echos of amplitudes that decay exponentially with a time constant T_2 would be observed. This sequence of pulses, which may be written $90° - \tau - (180° - 2\tau)_n$, is known as the Carr–Purcell sequence (Carr and Purcell 1954). In a simple modification to this sequence the $180°$ pulses are applied about the y'-axis of the rotating frame rather than about the x'-axis (corresponding to a phase shift of $90°$). This has two important effects: firstly the echoes all form along the same direction rather than alternately along the y'- and $-y'$-axes, and secondly any errors in the accuracy of the $180°$ pulses tend to cancel out rather than to add as they would do in the absence of the phase shift. This modification which was proposed by Meiboom and Gill (1958) results in what is now known as the Carr–Purcell–Meiboom–Gill (CPMG) sequence.

A proviso was mentioned above about diffusion. If there are imhomogeneities in the field B_0, diffusion of molecules will cause nuclei to experience different fields before and after the $180°$ pulses. Each $180°$ refocusing pulse relies on the precession frequency of the nuclei remaining constant, and therefore the effects of diffusion will be to reduce the amplitudes of the echoes. These effects can be minimized by decreasing the time interval 2τ between consecutive $180°$ pulses and, for example, if 2τ is made as low as 1 ms it should generally be possible to ignore the effects of diffusion. However, it is interesting to note that one can actually enhance the effects of diffusion by superimposing well-defined gradients upon the field B_0. Spin-echo experiments performed under such conditions enable diffusion constants to be measured.

In addition to providing measurements of spin—spin relaxation times, spin echoes have a number of additional applications. For example, Brown *et al.*(1977*a*), in their ^1H n.m.r. studies of red cells, collect free-induction decays following a

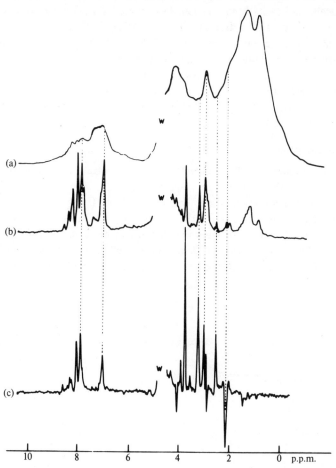

Fig. 6.11. ^1H n.m.r. spectra of red blood cells illustrating the effects of using the spin-echo pulse sequence. (a) a normal Fourier transform spectrum, in which the broad haemoglobin signals dominate. (b) a spectrum obtained using a $90°-\tau-180°-\tau$ spin-echo sequence, with $\tau = 20$ ms. This removes most of the haemoglobin signals. (c) same as (b), except that now $\tau = 60$ ms. The signals are even sharper, and some are inverted as a result of the phase modulation. (From Brown *et al.* 1977*a*).

$90° - \tau - 180° - \tau$ sequence. The free-induction decays that are observed constitute echoes in which the amplitudes of all the various signals are reduced according to their individual T_2 values. These values are much shorter for most of the haemoglobin resonances than for the metabolites, and therefore τ can be selected in such a way that the haemoglobin signals almost disappear while the metabolite signals remain. This sequence therefore provides a powerful means of spectral simplification (see Fig. 6.11).

When there is homonuclear spin–spin coupling (i.e. coupling between like nuclei, e.g. between two protons), echo formation becomes more complex but also more informative, as discussed by Brown and Campbell (1980) and in more detail by Rabenstein and Nakashima (1979). For example, a feature of doublets is that they become inverted if τ is equal to $1/J$, $3/2J$, $5/2J$, etc., while triplets and quartets have a different characteristic behaviour. This phase modulation of multiplet patterns explains why some of the signals in Figs. 3.6 and 6.11 are inverted, and it has a variety of interesting applications. For example, the effect can sometimes be used to observe selectively one or more multiplets from within a group of overlapping multiplets, thereby enhancing resolution and facilitating assignment of the spectra.

6.4. Signal intensities

The areas, or intensities, of the signals within a spectrum are proportional to the number of nuclei that contribute towards them, but are also dependent upon additional factors including (i) the effects of saturation (see § 7.2.4), (ii) the nuclear Overhauser effect (see § 6.6), (iii) the effects of finite pulse width (see § 7.2.5), and (iv) other spectrometer problems such as incorrect filter settings (see § 7.4.1). If we assume that these additional factors have negligible effects (or alternatively if the appropriate corrections for their effects are made), then the relative areas of the

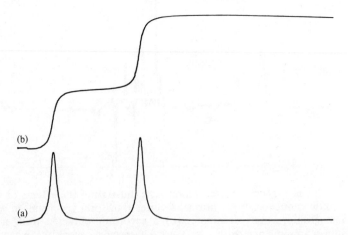

(b)

(a)

Fig. 6.12. (a) An n.m.r. spectrum, and (b) its integral, which measures the signal areas. The integral $\int g(\nu)d\nu$ of each signal is given by the height of the corresponding step in the integral curve.

signals within a spectrum will be equal to the relative numbers, and hence concentrations, of the nuclei that contribute towards them. Intensity measurements are therefore of great value in n.m.r.

The areas of n.m.r. signals are commonly measured by integration (see Fig. 6.12) using the computer. Alternatively, they can be estimated by plotting the spectra on graph paper and counting the squares underneath the peak, or by plotting the spectra, cutting out the peaks, and weighing. This assumes, of course, that the paper on which the peak is plotted is uniform. These latter methods may seem crude, but they avoid using valuable time on the computer, and for complicated spectra they may provide estimates of similar accuracy to those provided by the computer.

It is important to appreciate the difficulties associated with precise measurements of peak areas. One of the main problems is associated with the Lorentzian lineshape that characterizes many n.m.r. signals. Consider a Lorentzian signal of shape $g(\nu)$ given by

$$g(\nu) \propto \frac{T_2}{1 + 4\pi^2 (\nu - \nu_0)^2} \tag{6.17}$$

as shown by the solid curve in Fig. 6.13. The 'wings' of the signal stretch to $\pm\infty$, but in practice it is necessary to estimate the area by calculating the integral between certain finite limits, which we shall call $\nu_0 \pm \nu_I$. The measured integral I is then

$$I = \int_{\nu_0 - \nu_I}^{\nu_0 + \nu_I} g(\nu)\,d\nu$$

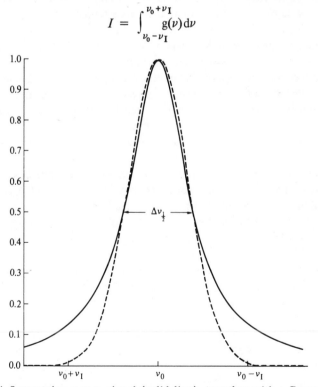

Fig. 6.13. A Lorentzian n.m.r. signal (solid line), together with a Gaussian (dotted curve) of the same width and height.

and it can be shown that

$$I = [\tan^{-1}\{2\pi(\nu_0 - \nu)T_2\}]_{\nu_0 - \nu_I}^{\nu_0 + \nu_I} \tag{6.18}$$

If $\Delta\nu_{1/2}$ is the width at half-height, we can conclude that 90 per cent of the total integral is obtained if ν_I is equal to 3.2 $\Delta\nu_{1/2}$, and 95 per cent is obtained if ν_I is equal to 6.5 $\Delta\nu_{1/2}$. Therefore, if a signal has a width of 10 Hz, an integral measured over a total of 130 Hz gives only 95 per cent of the true area. In order to obtain 98 per cent of the total area, the integral must be measured over about 300 Hz. In a high-resolution spectrum this can rarely be done because of the presence of other signals within the region.

The large amount of signal in the wings of a Lorentzian line is highlighted by comparison with the more familiar normal or Gaussian distribution shown by the dotted curve in Fig. 6.13. This curve has the same height and width at half-height as the Lorentzian, and it can be seen that the two curves are fairly similar in the central region but differ dramatically in the wings.

If it was known that an n.m.r. signal were truly Lorentzian, then its integral could be calculated from measurements of its height and width. However, the line-shape will often be perturbed by effects such as magnetic field inhomogeneities and unresolved spin–spin couplings. Because of these problems, together with additional uncertainties that may result from noise, sloping baselines, and neighbouring signals, we usually have to be content with uncertainties of ± 10 per cent in signal intensities. However, if we wish to compare the intensity of a given signal in two different spectra, changes of less than 5 per cent may be detectable in difference spectra obtained by subtracting one spectrum from the other. For such experiments it is essential to ensure that a control peak remains constant in the two spectra in order to confirm that there are no small intensity changes due to spectrometer instability. When comparing signal intensities in different spectra, it is often simpler and more accurate to compare peak heights rather than areas, because if the linewidth remains constant the area of the peak is directly proportional to its height.

The above discussion concerns the measurement of *relative* concentrations. In order to determine the *absolute* concentration of a compound it is necessary to adopt a calibration procedure whereby its signal intensity is compared with that from a known quantity of a reference compound. However, the calibration is more complicated for living systems than for solutions because it is difficult to assess the volume of cells or tissue that contributes to signal. Therefore special calibration procedures must be adopted, as for example in the studies of frog sartorius muscle described by Dawson *et al.* (1977*a*). The particular procedure that is used will depend upon the experimental conditions and on the type of preparation. A calibration system used for the quantitative measurement of metabolites in the rat heart is illustrated in Fig. 6.14. A ^{31}P n.m.r. spectrum is obtained under non-saturating conditions, and contains signals from a known concentration of inorganic phosphate and from the standard methylene diphosphonate which is contained within the annulus of a double-walled tube. The effective amount of methylene diphosphonate seen by the coil can be deduced from the relative peak areas. The

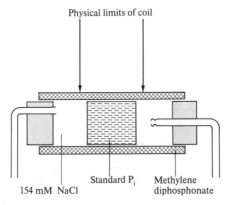

Fig. 6.14. A calibration system used for the quantification of metabolites in the rat heart. (From Garlick 1979).

inorganic phosphate is then replaced by a perfused heart which occupies approximately the same position within the tube. The absolute amounts of metabolites within the heart can be measured from a comparison of the areas of the heart metabolite signals with the peak area of the methylene diphosphonate together with the appropriate controls for saturation etc. These can then get converted to concentrations in terms of micromoles per gram wet weight. This method works provided that the heart and the inorganic phosphate standard are both fully enclosed by the coil. If either extended beyond the coil, there would be problems associated with the fact that different parts of the sample differ in their effectiveness at generating signal within the coil. The procedure adopted by Dawson *et al.* (1977*a*) overcomes this particular problem, but relies on assumptions that may not be valid for the perfused heart. In view of these difficulties, it is fortunate that for many   purposes the ratios of concentrations provide a sufficient basis for interpretation.

Finally, it should be pointed out that the term intensity is sometimes incorrectly used to indicate the height of a signal. It should be stressed that intensity refers to peak area, *not* to peak height.

6.5 Chemical exchange processes

N.m.r. spectra are sensitive to chemical exchange processes that take place under steady state or equilibrium conditions, and in this section we discuss some of the important aspects of chemical exchange measurements.

Let us consider a simple first-order equilibrium reaction

$$A \underset{k_B}{\overset{k_A}{\rightleftharpoons}} B ,$$

where k_A and k_B are the first-order rate constants for the reaction. We shall use this reaction to illustrate a number of different features of chemical exchange.

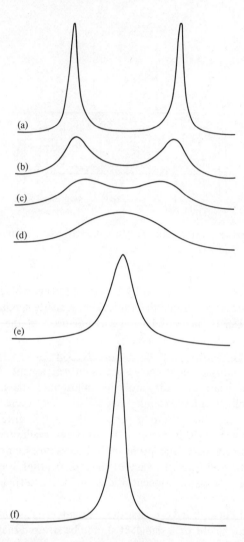

(a)

(b)

(c)

(d)

(e)

(f)

Fig. 6.15. Diagram illustrating the effects of chemical exchange between two species A and B of equal concentration. The exchange rate gradually increases on moving down from spectrum (a) to (f). (a) represents slow exchange, (b)–(e) represent various types of intermediate exchange, and (f) represents fast exchange. The actual rate constants are as follows: (a) $k = \pi(\nu_A - \nu_B)/10$; (b) $k = 3\pi(\nu_A - \nu_B)/10$; (c) $k = \pi(\nu_A - \nu_B)/2$; (d) $k = 4\pi(\nu_A - \nu_B)/5$; (e) $k = 2\pi(\nu_A - \nu_B)$; (f) $k = 4\pi(\nu_A - \nu_B)$.

Consider first two chemical species A and B of equal concentration which in the absence of exchange produce two narrow resonances at frequencies ν_A and ν_B. Since the concentrations are equal, we can write $k = k_A = k_B$. The effects of exchange on the observed spectra are shown in Fig. 6.15. It is convenient to distinguish between three ranges of exchange rates: slow in which $k \ll (\nu_A - \nu_B)$, fast in which $k \gg (\nu_A - \nu_B)$, and intermediate. When there is slow exchange between the two species, the resonances remain resolved but each is broadened by an amount $\Delta\nu = k/\pi$. This is equivalent to the lifetime broadening we considered in § 6.3.2; each nucleus only has a lifetime $\tau_m = 1/k$ in a given state and therefore there is an inherent uncertainty equal to k/π in the resonance frequencies.

In fast exchange the nuclei see an average of the A and B environments, and so we obtain a single signal at the average frequency. The width $\Delta\nu$ of this signal decreases as the exchange rate increases (by analogy with the 'motional narrowing' effect on T_2 that was discussed in § 6.3.2), and is given by

$$\Delta\nu = \pi(\nu_A - \nu_B)^2/2k. \tag{6.19}$$

As the exchange rate increases the averaging process becomes more complete and the line narrows.

In intermediate exchange the spectrum is more complex, but the lineshape for a given exchange rate can be predicted from the theory. The separation of the two signals in slow and intermediate exchange is given by

$$\delta\nu = \{(\nu_A - \nu_B)^2 - 2k^2/\pi^2\}^{1/2}. \tag{6.20}$$

Therefore the lines gradually merge as the exchange rate increases.

A rather interesting example of the above effects is illustrated in Fig. 6.16. This shows a ^{31}P n.m.r. spectrum obtained at close to 0°C from a solution containing 5 mM ATP, and just over 2.5 mM MgCl$_2$. The MgCl$_2$ concentration was chosen so that half of the ATP was free, and half was complexed to Mg^{2+} ions. Chemical exchange takes place between free ATP and MgATP, and the chemical shifts of

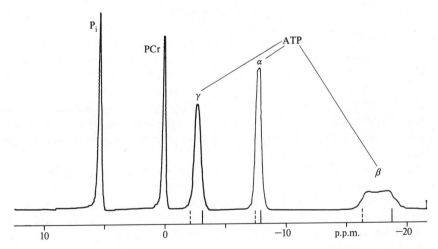

Fig. 6.16. ^{31}P n.m.r. spectrum at 73.8 MHz of a solution containing inorganic phosphate (P$_i$), phosphocreatine (PCr), and ATP at 0°C. Half of the ATP is complexed to Mg^{2+} ions, and half is free, and the spectrum shows the effects of chemical exchange between the two ATP species. The chemical shifts of free ATP and MgATP are indicated by the solid and dashed lines respectively. (From Pitkethly and Gadian, unpublished observations.)

these two species are indicated by the solid and dashed lines respectively. It is clear from the spectrum that free ATP and Mg-ATP exchange at such a rate that the β-phosphate signal is in intermediate exchange, whereas the α- and γ-phosphate signals are in relatively fast exchange. This illustrates the important point that the exchange is denoted slow, intermediate, or fast on the basis of its rate relative to the frequency separation of the two signals. From the spectrum it can be deduced that, under the conditions of the experiment, the dissociation of Mg^{2+} from ATP has a rate constant of about $380\,s^{-1}$.

When the concentrations of the two species are not equal, the calculations become much more difficult (see Pople, Schneider, and Bernstein 1959). However, the qualitative features of fast and slow exchange remain similar to those outlined above. In particular, when exchange is slow enough to ensure that both signals are clearly resolved, lifetime broadening of the signals may be observed. The more dilute species (B) has a shorter lifetime than A, and hence its signal broadens more extensively than the signal from A under conditions of slow exchange. When there is fast exchange the observed spectrum consists of a single resonance, the frequency of which is determined by the relative concentrations of the two species:

$$\nu = \frac{[A]\,\nu_A + [B]\,\nu_B}{[A] + [B]} \tag{6.21}$$

Fast exchange is encountered whenever we use the frequency dependence of a resonance to determine pH. Consider, for example, inorganic phosphate, which exists mainly as HPO_4^{2-} and $H_2PO_4^-$ at around neutral pH. The equilibrium between these species can be written

$$H^+ + HPO_4^{2-} \rightleftharpoons H_2PO_4^-.$$

In the absence of chemical exchange, the two species would give rise to two resonances separated from each other by about 2.4 p.p.m. In solution, however, HOP_4^{2-} and $H_2PO_4^-$ exchange with each other rapidly (at about $10^9-10^{10}\,s^{-1}$), and as a result a single resonance is observed, of frequency ν given by eqn (6.21), where A and B represent the two ionization states of inorganic phosphate. Since the relative concentrations of HPO_4^{2-} and $H_2PO_4^-$ are determined by $[H^+]$, the frequency of the observed signal measured as a function of pH gives a 'standard' pH curve (see § 2.2).

In addition to affecting chemical shifts and linewidths and shapes, exchange also tends to average the spin–lattice relaxation of the two resonances. In general, this will cause the observed spin–lattice relaxation to be non-exponential. However, when species A has a far greater concentration than species B, the measured spin–lattice relaxation of A remains exponential, with a time constant $T_{1\,obs}$ given by (Luz and Meiboom 1964)

$$1/T_{1\,obs} = 1/T_{1A} + f/\tau_B + T_{1B} \tag{6.22}$$

where T_{1A} and T_{1B} are the inherent T_1 values of A and B in the absence of exchange, f is the ratio of B to A, and τ_B is the lifetime of the B species. In fast exchange τ_B is much smaller than T_{1B}, and therefore

$$1/T_{1\,obs} = 1/T_{1A} + f/T_{1B}. \qquad (6.23)$$

This equation describes complete averaging of the relaxation rates.

Our discussion so far has assumed that the two signals are very narrow in comparison with their frequency separation. Often, this is not true; in fact, we frequently encounter the situation in which a ligand is exchanging between solution and a bound state. The bound ligand generally produces a much broader signal than the free form, and under these conditions the equation describing the observed lineshape can become very complex. The equation can be simplified if it is assumed that the bound ligand is at a much lower concentration than the free form (see Swift and Connick 1962), and it can be simplified even further if there is a very small frequency difference between the two signals. If these assumptions are valid, the observed spin–spin relaxation time of the free form is given by

$$1/T_{2\,obs} = 1/T_{2A} + f/\tau_B + T_{2B} \qquad (6.24)$$

where A is the free and B is the bound form of the ligand. This equation is exactly analogous to the expression for T_1 given above.

6.5.1. The determination of exchange rates

If exchange rates are very slow, i.e. in the range of minutes to days, they can be detected by applying a perturbation to the system and watching the return to equilibrium. For example, if a protein is dissolved in D_2O, the rate at which the tryptophan indole NH protons exchange with the solvent can be detected by monitoring the rate at which the intensity of the corresponding ^1H signals decline. We would expect the rate of exchange to be faster for tryptophan residues that are close to the surface of the molecule than for those residues that are buried in the interior of the protein. Such experiments therefore provide a valuable aid to the determination of protein conformation (Dobson 1977).

Exchange rates in the range $0.1-1\,s^{-1}$ can often be detected by their effects on T_1 and phenomena such as saturation that are related to T_1 (see Appendix 6.1). However, exchange rates in this range are generally too slow to have significant effects on the widths and shapes of signals obtained from biological molecules. Faster rates of the order of $100\,s^{-1}$ do have dramatic effects on the spectra, as we have observed in the Mg-ATP example described above. Even faster exchange rates can be measured when paramagnetic probes are used to enhance relaxation (see Dwek 1973).

6.6. The nuclear Overhauser effect

We have seen in § 6.3.4 that the relaxation for many nuclei is dominated by dipole–dipole interactions between neighbouring nuclear spins. For example, protonated ^{13}C nuclei are relaxed largely through this type of interaction with neighbouring protons. Under normal conditions the relative populations of the protons in their two spin states are given by the Boltzmann distribution (see § 5.2.4). It can be shown that if this population distribution is altered, for example by saturating the proton spins, the dipole—dipole interaction with the neighbouring ^{13}C nuclei causes a change to take place in the relative populations of the ^{13}C spin states. This results in a change in the steady state magnetization of the ^{13}C nuclei, and is known as the nuclear Overhauser effect (for more detail, see Campbell and Dobson, 1979).

It can be shown that saturation of the ^{1}H resonances (which corresponds to the populations of the spin states being equalised) can lead to an enhancement in the intensity of the *observed* signals given by the expression

$$\frac{\text{signal intensity (^{1}H saturated)}}{\text{signal intensity (no saturation)}} = 1 + \frac{\gamma_s}{2\gamma_0} \qquad (6.25)$$

where γ_s and γ_0 are the magnetogyric ratios of the saturated and observed nuclei respectively. However, this equation is only valid if the spin—lattice relaxation of the observed nuclei is caused entirely by dipole—dipole coupling, and also if the motion is sufficiently rapid for the condition $\omega_0\tau_c \ll 1$ to be satisfied. If either of these criteria is not fulfilled, the enhancement is decreased; indeed, when the motion is slow, the observed signal intensity may actually decrease rather than increase.

In ^{13}C n.m.r. experiments the nuclear Overhauser effect often produces a significant intensity enhancement, and therefore proton irradiation is usually applied in order to obtain this important improvement in the signal-to-noise ratio. The degree of enhancement provides an excellent indication of the contribution to relaxation by dipolar coupling, and can be a powerful aid to conformational analysis because the magnitude of the dipolar coupling is strongly distance dependent (proportional to $1/r^6$). It should be noted that proton irradiation can also simplify spectra and further improve the signal-to-noise ratio by means of spin decoupling (see below and § 6.2.4).

The nuclear Overhauser effect is particularly useful in ^{15}N n.m.r. This is because the ^{15}N nucleus has a small magnetogyric ratio, and from eqn (6.25) it can be seen that this leads to a large enhancement. It is interesting to note that the magnetogyric ratio of the ^{15}N nucleus is negative, and therefore in general the enhancement factor is also negative. It is for this reason that many of the signals in ^{15}N n.m.r. spectra appear inverted.

The nuclear Overhauser enhancement for ^{31}P nuclei is usually smaller than for ^{13}C nuclei, partly because the magnetogyric ratio for ^{31}P nuclei is larger and partly because ^{31}P nuclei are rarely bonded directly to protons. It is therefore not so common to employ proton irradiation in biological ^{31}P n.m.r. experiments for the intensity gains are often outweighed by the additional controls that are required in order to relate peak intensities to concentrations. Furthermore, the ^{31}P–^{1}H spin—spin coupling constants are usually so small that there is little to be gained by spin decoupling.

Effective saturation is in general achieved provided that the condition $\gamma B_2 \gg (T_1 T_2)^{-1/2}$ is satisfied, where B_2 is the strength of the irradiating field and T_1 and T_2 are the relaxation times of the resonances being saturated. This is a weaker requirement than for spin decoupling (see § 6.2.4), and therefore the irradiation used in spin-decoupling experiments is always sufficiently powerful to ensure that the nuclear Overhauser effect is obtained. Continuous double irradiation of the ^{1}H signals in ^{13}C n.m.r. experiments can therefore result in (i) collapse of multiplets which simplifies spectra and also enhances signal intensities as two or more lines are collapsed into one, and (ii) additional enhancements of signal intensities by factors of up to 3 due to the nuclear Overhauser effect. 'Double resonance' experiments of this type are discussed further in § 7.2.6.

Appendix 6.1. Saturation transfer

Saturation transfer is an n.m.r. technique that has proved particularly useful for monitoring the steady state kinetics of enzyme-catalysed reactions *in vivo*. In this appendix we briefly outline the theory of the method, which has been discussed in depth by Forsen and Hoffman (1963, 1964).

Consider the equilibrium reaction

$$A \underset{k_B}{\overset{k_A}{\rightleftharpoons}} B,$$

where k_A and k_B are the first-order rate constants for the forward and back reactions respectively. Suppose that the equilibrium values of the magnetization of A and B are M_{0A} and M_{0B} respectively, while the z components of magnetization at any given time are M_A and M_B. In the absence of exchange between A and B, M_A would return to its equilibrium value M_{0A} with a time constant T_{1A}, and we could write

$$\frac{dM_A}{dt} = \frac{M_{0A} - M_A}{T_{1A}}$$

In the presence of exchange, this equation must be modified as A loses magnetization through conversion to B and gains magnetization from the reverse process. The full equation under these conditions is

$$\frac{dM_A}{dt} = \frac{M_{0A} - M_A}{T_{1A}} - k_A M_A + k_B M_B \qquad (6.26)$$

Suppose now that the resonance of B is selectively saturated so that $M_B = 0$. Equation (6.26) then becomes

$$\frac{dM_A}{dt} = \frac{M_{0A} - M_A}{T_{1A}} - k_A M_A. \qquad (6.27)$$

In the steady state $dM_A/dt = 0$, and therefore

$$0 = \frac{M_{0A} - M_A}{T_{1A}} - k_A M_A$$

from which

$$M_A = \frac{M_0}{1 + k_A T_{1A}} \qquad (6.28)$$

i.e. the magnetization of A in the steady state is no longer equal to M_{0A}, but is equal to M_{0A} reduced by a factor $1 + k_A T_{1A}$.

Consider experiments in which radiofrequency pulses are applied every t seconds. If t is much greater than T_{1A}, the magnetization of A reaches a steady state prior to application of each pulse, and so a comparison of the observed signal intensity in the presence and absence of selective irradiation of B yields the magnitude of $1 + k_A T_{1A}$. In order to determine k_A, it is then necessary to measure T_{1A}. Unfortunately, T_{1A} cannot be measured from a conventional T_1 experiment because T_{1A} represents the inherent spin–lattice relaxation time of A that would be observed in the absence of chemical exchange. Any measurement made in the presence of exchange must necessarily include a contribution from the relaxation of B. This contribution can be evaluated using the theory developed by Forsen and Hoffman, but there is an alternative simpler method of obtaining the required information.

The full solution to eqn (6.26) shows that, when B is saturated, M_A returns to its steady state value $M_{0A}/(1 + k_A T_{1A})$ with a time constant T_{1eff} given by $1/T_{1eff} = 1/T_{1A} + k_A$. Therefore, a T_1 measurement made in the presence of selective irradiation of B enables k_A and T_{1A} to be determined from the simultaneous equations

$$M_A = M_{0A}/(1 + k_A T_{1A})$$

$$\frac{1}{T_{1eff}} = \frac{1}{T_{1A}} + k_A .$$

Similar measurements can be made for the determination of k_B and T_{1B}, and experimental observations on the reaction catalysed by creatine kinase are described in Chapter 3. It should be pointed out that several types of experimental procedure are available for determining the rate constants (see Campbell, Dobson, Ratcliffe, and Williams 1978; Nunnally and Hollis 1979; Gadian *et al*. 1981).

7

Spectrometer design and operation

In this chapter we describe the design, logic, and operation of a typical Fourier transform n.m.r. spectrometer. The aims are to provide the reader with an understanding of how an n.m.r. spectrometer works and to explain some of the common operational procedures that are involved in performing an experiment.

We saw in § 1.7 that an n.m.r. spectrometer can be divided into a number of units which were described as (i) the magnet, (ii) the probe, (iii) the transmitter, (iv) the receiver, and (v) the computer. In this chapter we discuss in some detail each of these units, except for the probe which has a chapter all to itself. We discuss how these units interact with each other, and explain how to adjust each of them for optimal operation of the spectrometer.

7.1 The magnet

7.1.1. The characteristics of the magnetic field

The capability of an n.m.r. spectrometer is critically dependent upon the magnitude and homogeneity of the static magnetic field and on the physical dimensions of the magnet. The magnet may be one of three different types: a permanent magnet, an iron-cored electromagnet, or a superconducting magnet. Of these, only superconducting magnets generate fields significantly higher than 2.3 T (corresponding to a frequency of 100 MHz for ^{1}H n.m.r.) and nowadays are the type most commonly employed for high-resolution biological n.m.r. A distinction between electromagnets and superconducting magnets is that the direction of the field is generally different in the two types (see Fig. 7.1). This affects probe design (see Chapter 8) and can affect chemical shift measurements when using external standards (see § 6.1.2).

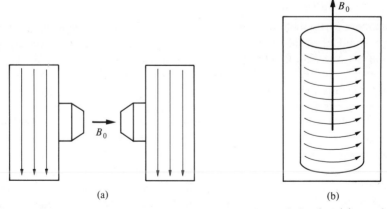

(a)   (b)

Fig. 7.1. Schematic diagram illustrating the direction of B_0 for (a) an electromagnet, and (b) a superconducting magnet. The light arrows represent the circulating electric currents that generate the field.

We saw in § 1.7 that a critical feature of any magnet designed for high-resolution n.m.r. studies is that the field should be extremely homogeneous over the sample volume. If linewidths of less than 1 Hz are required at frequencies of 300 MHz, then the field must be homogeneous to the remarkable value of close to 1 part in 10^9 over the sample volume. The basic homogeneity of any magnet is far worse than this, and it is necessary to use two techniques for accomplishing the required homogeneity. The first is to compensate for inhomogeneities in the field by adjustment of the shim coils (see § 7.1.3), each of which produces its own characteristic field profile. The second is to spin the sample at a rate sufficiently rapid to average out most of the residual inhomogeneities that may remain. Typical spinning speeds are about 20 Hz. In general the homogeneity requirements for studies of living systems are not as stringent as for high-resolution studies of solutions of macromolecules; often 1 part in 10^7 is perfectly satisfactory. As a result it may not be necessary to spin the sample, which is fortunate as the spinning of living systems might not be feasible.

Wide-bore magnets, with cylindrical bores of diameter 10 cm or more, have recently become available, and these have greatly extended the range of feasible experiments. For example, it is now possible to fit live rats into the bore of magnets of field strength 4.3 T, human limbs into magnets of field strengths 2 T, and human beings into magnets with lower field strengths. Of course, spectra can still only be obtained from those regions of the sample that experience a sufficiently homogeneous field; this consideration limits the useful, or active, volume of the magnet.

Another essential feature of the magnet is that the field should drift by a negligible amount during the course of an experiment. Any fluctuations in the field would generate corresponding fluctuations in the resonance frequencies, resulting in a broadening of the spectral lines. Some modern superconducting magnets are extremely stable and do not require any additional stabilizing devices. However, for many spectrometers it is necessary to employ a 'field-frequency' lock which causes the field to 'lock in' to a specific value.

7.1.2. The field-frequency lock

The purpose of the field-frequency lock is to stabilize the static magnetic field of the spectrometer so that resonance positions do not change during the course of an experiment. The field-frequency lock constitutes an additional channel within the spectrometer; in fact it can almost be regarded as a second spectrometer. Often, it employs the ^{2}D n.m.r. signal from a partially or totally deuterated solvent.

If a sample is dissolved in 90 per cent H_2O/10 per cent D_2O, a large ^{2}D signal can be detected which has a resonance frequency proportional to the field experienced by the nuclei. Therefore, if we can ensure that the ^{2}D resonance frequency remains constant, the field must also remain constant and the purpose of the lock has been achieved. The basic principle is to collect a ^{2}D signal as shown in Fig. 7.2; this is not the conventional absorption signal that we are accustomed to, but the dispersion mode signal that can also be obtained. The dispersion mode signal goes through zero at resonance and deviates from zero in the vicinity of resonance. Suppose that the field is such that the ^{2}D signal is precisely at resonance, i.e. its fre-

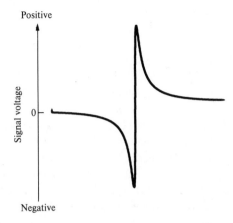

Fig. 7.2. The dispersion mode signal that is used for field-frequency locking.

quency is exactly equal to the frequency of the applied radiofrequency field. If the field changes slightly the observed signal will deviate from zero, and the resulting voltage causes a field to be developed that counteracts the field change that has taken place. This negative feedback system ensures that the deuterium signal is always at resonance and hence that the field strength of the spectrometer remains constant.

The above argument assumes that the frequency of the applied field B_1 remains precisely constant, i.e. we have considered instabilities that may exist in the field B_0 but we have not taken into account any instabilities there may be in the frequency generator. However, it can be shown that as long as the ^2D frequency is derived from the same source as the radiofrequency used for the detection of the required spectrum, the lock will compensate in the correct way for any instabilities in this source.

7.1.3. Adjustment of the shim coils

The shim coils are fitted within the bore of the magnet. Each is designed in such a way that passage of electrical current through it generates a well-defined field gradient. For example, the z coil produces a field that varies linearly in the z direction, the z^2 coil produces a field proportional to z^2, etc. Thus, for example, if the inhomogeneity in the basic field is largely due to the existence of a small linear gradient in the z direction, this can be 'shimmed out' using the z shim coil.

The procedure for optimizing the field homogeneity will vary according to the signal that is being observed and the method whereby it is detected. The signal should be intense enough to be easily detectable in a single scan, which eliminates many possiblities. The field-frequency lock signal, which is often a ^2D n.m.r. signal from D_2O added to the sample, is generally employed for shimming in routine n.m.r. experiments. (Although ^2D has a spin of 1, D_2O nevertheless produces a ^2D signal that is narrow and hence can be used for shimming.) However, for studies of living systems a lock may be unnecessary, and furthermore it may be unacceptable to add large amounts of D_2O or any other material to the sample. Under such cir-cumstances it may be possible to use the ^1H signal from the water within the sample for homogeneity adjustment as described by Ackerman *et al.* (1981). This method is thoroughly recommended, and it should be ensured that the facility is available on any spectrometer routinely used for studies of living systems.

Regardless of the method that is employed, the aim of shimming is always to ensure that the signal is as narrow as possible, consistent with the lineshape also being presentable. This is illustrated in Fig. 7.3. Figure 7.3(b) shows a broad signal, which we can regard as being the deuterium resonance from D_2O within a sample. Figure 7.3(a) shows the corresponding FID which decays rapidly (remember that the time constant of the decay is inversely proportional to the linewidth). We know that the intrinsic linewidth of the ^2D signal from D_2O is much narrower than that shown in Fig. 7.3(b), and we can therefore conclude that the observed width results from inhomogeneities in the field B_0. Improved shimming produces the results shown in Figs. 7.3(c) and 7.3(d). It should be noted that the signal in Fig. 7.3(d) contains a broad and a narrow component, signifying that the field is homogeneous in one part of the sample but inhomogeneous in another part (probably the

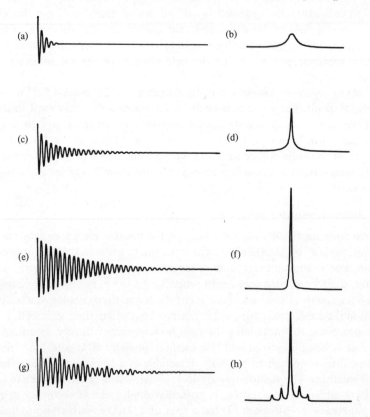

Fig. 7.3. Free induction decays and their Fourier transforms, observed at various stages of a field shimming procedure. For further details, see text.

extremities of the sample, either radial or axial). The corresponding FID has a rapidly decaying and a slowly decaying component, corresponding to the broad and narrow components respectively of the signal. Additional adjustments of the shim coils produces the results shown in Figs. 7.3(e) and 7.3(f). The broad 'wings' of Fig. 7.3(d) have now disappeared, and so there is more intensity in the narrow component of the signal with a consequent improvement in the signal-to-noise ratio. Finally, we show in Figs. 7.3(g) and 7.3(h) what can happen with a spinning sample in an inhomogeneous field. Rotation of the sample in an inhomogeneous field causes the nuclei within the sample to experience a field which is modulated at a frequency equal to the frequency of rotation. As a result, small artefactual peaks appear on both sides of the actual signal at separations equal to the frequency of rotation. Correct adjustment of the appropriate shim coils should cause the intensity of these 'spinning sidebands' to fall to about 1 per cent or less of the main signal intensity. If there is a solvent peak which is very large in comparison with other signals within a spectrum, care must be taken to ensure that some of these signals are not simply spinning sidebands of the solvent peak. This can be checked by altering the spinning speed; this will change the frequencies of the spinning sidebands without affecting the frequencies of the genuine resonances. The theory of spinning sidebands is dealt with in depth by Hoult (1978) who points out that spinning sidebands can also be generated as a result of inhomogeneities in the radio-frequency field B_1.

7.1.4 The choice of magnet

It is generally considered desirable to perform experiments at the highest magnetic field that is available, for in general an increase in field strength will increase both sensitivity and resolution. A high field is certainly desirable for solution studies, and indeed for almost all experiments utilising relatively small amounts of material (less than $1-2$ ml). However, for studies of relatively large living systems, there are additional criteria that must be considered.

In particular, the spectrometer must be versatile with respect to sample size and shape. This means that the magnet must have a wide bore, and must produce a field that is homogeneous over a large volume. However, magnets of very high field strengths tend to have narrow bores and relatively small regions of field homogeneity. This is because of problems associated with the cost, and in extreme cases with the technical feasibility of building physically large magnets of high field strength. (It should be noted that the cost of the magnet represents a significant, and in some cases major percentage of the total cost of a n.m.r. spectrometer).

There are a number of additional reasons why high field strengths are not necessarily desirable for studies of living systems. Firstly, it is difficult to design large radiofrequency coils that perform well at high frequencies (see § 8.7). Secondly, there are problems associated with the penetration of the radiofrequency field into very large samples, and this effect increases as the frequency increases (see § 8.5.3). Finally, the effects of chemical exchange, and of chemical shift anisotropy (particularly for ^{31}P; see § 6.3.4) can reduce the gain in resolution that might be anticipated on increasing the field strength.

Taking all of these factors into consideration, it is possible to draw up general guidelines as to the most suitable field strengths for various types of study. For example, it seems that a suitable field for whole-body ^{1}H imaging is about 0.1 T (see § 4.3). Suitable field strengths for ^{31}P n.m.r. are approximately 2 T for human studies, 4–5 T for whole tissues and organs, and about 8 T for studies of cellular suspensions. For corresponding ^{13}C studies of living systems, rather higher field strengths are probably desirable. Finally, we should reiterate that for studies of relatively small amounts of solution, the most suitable field will in general be the highest that is available.

7.2. The transmitter

7.2.1. The basic design

The function of the transmitter is to irradiate the n.m.r. sample with radiofrequency fields of the appropriate frequency, power, and timing. The transmitter contains a frequency generator, a 'gate' which switches the transmission on and off at the required times, and a power amplifier which boosts the radiofrequency power to the values that are required in Fourier transform n.m.r. (see Fig. 7.4). The timing of

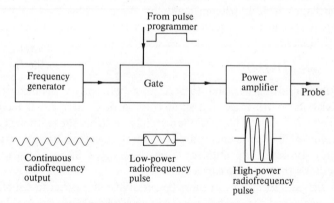

Fig. 7.4. A simple block diagram of the spectrometer transmitter. Beneath each stage is shown the type of output that is generated.

the radiofrequency pulses that are used for data accumulation, spin decoupling etc. is controlled by a part of the spectrometer known as the pulse programmer. Usually, this constitutes part of the interface between the computer and the rest of the spectrometer, and the pulse timings can be set simply by entering the required values into the computer.

In a typical simple experiment a spectrum is accumulated over a given period of time by applying radiofrequency pulses at regular intervals, which we shall call the pulse interval T_p. The selection of the irradiating frequency is usually fairly straightforward, but the precise value relative to the frequencies of the resonances will be affected by certain features of the receiver design (see § 7.3). The other important variables that must be considered when setting up the transmitter for this simple experiment are the duration of the pulses (commonly known as the pulse width) and the pulse interval. Great care has to be taken in selecting their values, because any errors can result in dramatic losses in the signal-to-noise ratio. Let us now consider how to choose the pulse widths and intervals that provide optimal signal-to-noise ratios.

7.2.2 Theory governing the choice of pulse width and interval

We recall from § 5.3.1 that a radiofrequency pulse of duration or width t_p tilts the magnetization through an angle θ given by

$$\theta = \gamma B_1 t_p. \tag{7.1}$$

θ is known as the pulse angle, and this equation shows that it is linearly related to the pulse width. If we assume that the magnetization prior to the pulse is M_0 and is directed along the z axis, then the pulse generates a component M_{xy} in the xy plane given by

$$M_{xy} = M_0 \sin \theta$$
$$= M_0 \sin \gamma B_1 t_p. \tag{7.2}$$

A $90°$ pulse is one that tilts the magnetization totally into the xy plane, and is of length t_p^{90} where

$$\gamma B_1 t_p^{90} = \pi/2. \tag{7.3}$$

It is the magnetization M_{xy} that is detected in the n.m.r. experiment.

We saw in § 5.5.3 that in general the setting of the pulse interval involves a compromise. If the interval is too small, there is a danger of saturating the signals, with a consequent loss in signal because the magnetization prior to each pulse is less then M_0. If the interval is too large, then in a given period of time too few scans are collected, and since the signal-to-noise ratio increases according to the square root of the number of scans this also adversely affects the signal-to-noise ratio.

It is not unduly difficult to calculate how the signal-to-noise ratio obtained in a given time varies with pulse angle and interval; the relevant equation is given by Becker, Ferretti, and Gambhir (1979) following the analysis of Ernst and Anderson (1966), and the results are shown in Fig. 7.5. From this diagram, it can be seen that if we wish to apply $90°$ pulses then the optimal signal-to-noise ratio is obtained if the pulses are applied at intervals of $1.25\,T_1$, where T_1 is the spin–lattice relaxation time of the signal of interest. However, it is even better to apply shorter pulses rather more rapidly. For example, $45°$ pulses applied every $0.35\,T_1$ will generate a 10 per cent higher signal-to-noise ratio in a given period of time than $90°$ pulses applied every $1.25\,T_1$. The optimum angle for a given pulse interval is known as the Ernst angle α; it is given by the expression

$$\cos \alpha = \exp\left(-T_p/T_1\right) \tag{7.4}$$

and is plotted in Fig. 7.6.

The curves plotted in Fig. 7.5 suggest that the optimal approach might be to use

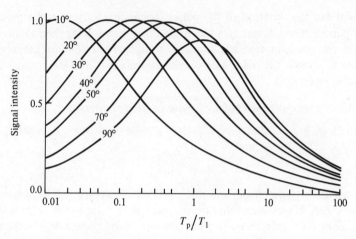

Fig. 7.5. The signal-to-noise ratio expressed as a function of the interval between successive radiofrequency pulses, for various pulse angles. (From Shaw 1976.)

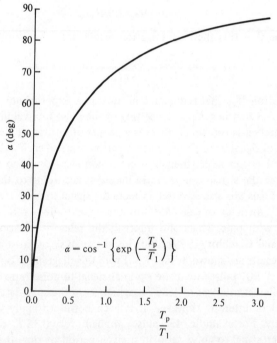

$$\alpha = \cos^{-1}\left\{\exp\left(-\frac{T_p}{T_1}\right)\right\}$$

Fig. 7.6. The Ernst angle as a function of pulse interval. (From Shaw 1976.)

very small pulse angles and correspondingly small pulse intervals. A second advantage of this procedure is that any errors arising from finite pulse width (see § 7.2.5) are minimized. However, this is not necessarily the best or most practical approach for a number of reasons. Firstly, it is necessary to collect the FID following each pulse over a period of about $3-4\,T_2^*$; this period of data collection (known as the acquisition time) sets a practical lower limit to the pulse interval. Thus if T_2^* is 30 ms and T_1 is 1 s, the acquisition time may be about 100 ms, and the most rapid application of pulses is every $0.1\,T_1$; from eqn (7.4) the optimum pulse angle under such circumstances would be about $25°$. Secondly, there may be instrumental problems (i) in obtaining accurate timing of the pulse lengths if these are very short and (ii) in ensuring that there are no adverse effects resulting from the addition in the computer of a very large number of FIDs. However, these problems should not be too serious if the spectrometer is well designed. Another factor that it may be necessary to consider is that one might wish to apply a second irradiation frequency during periods between the acquisition time, as shown in Fig. 7.9, and this may set further lower limits on the feasible pulse interval. A safe compromise that would suit many situations would be to apply $45°$ pulses every 0.35 T_1.

7.2.3. Measurements of the 90° pulse width

Figures 7.5 and 7.6 give the signal-to-noise ratios that can be expected for a given pulse interval and *angle*. However, the variable that one actually adjusts is the pulse *width*, and it is therefore essential to know the proportionality constant relating pulse angle to pulse width (see eqn (7.1)). In practice, this is most easily done by measuring the length of a $90°$ or $180°$ pulse.

For a conventional spectrometer that always uses the same probe it is quite likely that the $90°$ pulse length will remain fairly constant from one experiment to another. However, it is often wise to check the $90°$ pulse length before starting an experiment, particularly if its value should be known accurately (as, for example, when carrying out T_1 and T_2 measurements). This is because the B_1 field, and hence the $90°$ pulse length, is dependent upon each of the following factors: (i) the output from the power amplifier, which should in fact remain constant; (ii) the nature of the probe, and the quality of its tuning and matching (see § 8.6); (iii) the nature of the sample, in particular its size and electrical conductivity (see § 8.5). If it is felt that the $90°$ pulse length should be checked the following procedure should be adopted, bearing in mind that we cannot for example use ^{1}H signals to check the $90°$ pulse for ^{31}P; in this case ^{31}P signals must be used.

A sample should be chosen that has a resonance sufficiently strong to be detected in a single scan. Preferably, this should be the actual experimental sample, but if none of the signals is strong enough a test sample should be chosen that affects the probe tuning and matching in exactly the same way as the experimental sample. This ensures that the B_1 field is the same for both samples and therefore that the $90°$ pulse length for the test sample is the same as that for the experimental sample. The pulse interval should be set at a value greater than $4\,T_1$, where T_1 is the spin–lattice relaxation time of the resonance to be detected; this ensures that there is no significant saturation. If the T_1 is totally unknown, one should play safe by choosing a very long pulse interval. The pulse width is gradually increased, and for each width a single FID is collected and Fourier transformed. On displaying the resulting signals, a sinusoidal variation of signal amplitude with pulse width should be observed, as predicted by eqn (7.2) and shown in Fig. 7.7. The $90°$ pulse

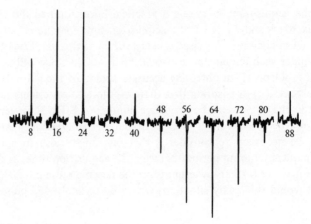

Fig. 7.7. Signal amplitude from a single scan, plotted as a function of pulse width. The pulse widths corresponding to each signal are given in μs.

length is, from eqn (7.2), the length that generates maximum signal, and from Fig. 7.7 is seen to be about 20 μs. A perfect 180° pulse is one that inverts the magnetization; M_{xy} following such a pulse is zero. The length of a 180° pulse is twice the length of a 90° pulse, and so the 90° pulse length can be measured by looking for the pulse length that gives zero signal and dividing by 2. Unfortunately, however, the effects of increasing pulse width often differ from the predicted behaviour as a result of imperfections in the B_1 field; if the B_1 field is not homogeneous over the sample volume, a 90° pulse for one part of the sample may only be a 45° pulse for another part. Deviations from predicted behaviour may be particularly evident in the region where a null signal should be obtained; a signal that is similar in shape to the dispersion mode signal is often observed. This is caused by the combined effects of B_0 and B_1 inhomogeneities. In addition, the signal following a 270° pulse, which should be equal and opposite to the signal following a 90° pulse, may be of much lower intensity than predicted as a result of B_1 inhomogeneity. When this is the case, care must be taken when interpreting measurements that rely on the accuracy of the 90° or 180° pulse.

7.2.4. T_1 effects – saturation factors

The relationships plotted in Figs. 7.5 and 7.6 express the pulse interval in units of T_1. Therefore if these plots are to be used to optimize pulsing conditions, it is necessary first to measure the approximate T_1 values of the resonances (see § 6.3.6). A slightly different approach, which is rather more empirical and less time consuming, is to decide to apply pulses of a given angle, say 45°, and to determine by trial and error the optimum interval for this particular angle. However, regardless of the approach that is used, if there are several resonances with different T_1 values then it is difficult to decide which of these to consider when choosing the optimal conditions; optimizing the signal-to-noise ratio for one resonance will certainly not optimize the signal-to-noise ratio for another.

Having selected a suitable pulse angle and interval, it will be necessary to perform careful controls if we wish to relate the observed peak areas to concentrations. The reason for this is that the relatively rapid pulsing that is required for optimal signal-to-noise ratios will cause partial saturation of the resonances, the extent of which will depend on their T_1 values. Consequently, before converting relative peak areas to relative concentrations, it is first necessary to multiply each area by a correction factor, termed its saturation factor. The saturation factor for any peak is equal to its area measured under non-saturating conditions divided by its area measured under the conditions of the experiment. This factor can probably be determined from the procedure that was carried out when optimizing the pulse interval. Suppose, for example, that we are performing an experiment in which $45°$ pulses are being applied at intervals of 1 s. It may have been found in setting up these pulsing conditions that an interval of 8 s between $45°$ pulses was sufficiently long to ensure that there was no significant saturation of any of the resonances. The saturation factor for a given peak is then simply the ratio of its area using 8 s and 1 s intervals.

7.2.5. *The effective bandwidth of radiofrequency pulses*

When considering the effects of a radiofrequency pulse, we have assumed so far that its frequency is equal to the resonance frequency of the nuclei. However, for spectra containing a large number of chemically shifted signals, this assumption can only be valid for at most one of the resonances. We must therefore consider how effective a pulse is when its frequency (which we now define as ν_1) differs from the resonance frequency ν_0 of the nuclei. In particular, we need to know whether all the resonances within a spectrum are excited equally by application of a pulse. If they are not, then the measured intensities of the signals will not be truly representative of the intensities that should be obtained.

In § 5.5.1 we briefly considered the effects of finite pulse width. In particular, it was noted that, as the pulse width increases, we might expect the effective bandwidth of the pulse to decrease. However, it was also pointed out that the precise frequency *response* to a radiofrequency pulse differs slightly from the frequency distribution of the pulse itself[1]. In order to determine the frequency response, we must resort to calculations using the Bloch equations of the type described by Meakin and Jesson (1973). Some results from these calculations are illustrated in Fig. 7.8. Six simulated spectra are shown, each of which is the Fourier transform of the response to a single $90°$ pulse. Each spectrum corresponds to a different B_1 field strength and hence to a different value for the $90°$ pulse length t_p^{90} (remember that B_1 and t_p^{90} are related to each other through eqn (7.3)). The horizontal axis represents the difference between the resonance frequency and the frequency of the applied B_1 field; this difference $\nu_0 - \nu_1$ is commonly called the offset. It can be seen that, as the offset increases, the intensity of the corresponding signal declines and its phase changes. The effects become more marked as t_p^{90} increases, i.e. as B_1

---

[1] There would be no difference between the two if n.m.r. were a linear system, i.e. if the output from the system were linearly proportional to the input. This is certainly not the case for n.m.r., because the response to a $180°$ pulse is *not* double the response to a $90°$ pulse.

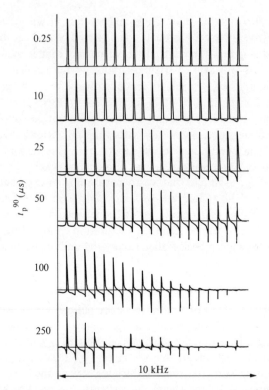

Fig. 7.8. The effect of finite pulse power (i.e. length of a 90° pulse) on signal phase and amplitude, plotted as a function of offset from 0 to 10 kHz. (From Meakin and Jesson 1973.)

decreases. Theory shows that if intensity errors of less than 2 per cent are required, the value of t_p^{90} must be less than $1/4\,\Delta$ where Δ is the maximum offset; this corresponds to a value of t_p^{90} of less than $50\,\mu s$ for an offset of 5 kHz.

Figure 7.8 refers to spectra obtained by application of 90° pulses. The intensity variations in the two lowest spectra of Fig. 7.8 are clearly unacceptable, but the situation can be improved considerably by applying smaller angle pulses. For example, a 45° pulse of length $50\,\mu s$ has about twice the effective bandwidth of a 90° pulse of length $100\,\mu s$, and a 22° pulse of length $25\,\mu s$ has a bandwidth about four times as great, i.e. there is an approximate inverse relationship between the duration of a pulse and its effective bandwidth, as might be expected from the arguments of § 5.5.1. Therefore, if the weakness of the B_1 field could conceivably lead to bandwidth problems, it would be preferable to apply small-angle pulses at intervals that can be determined by reference to eqn (7.4).

7.2.6. *Double resonance and solvent suppression*

In addition to the radiofrequency pulses that are used for routine spectral accumulation, it is often useful, or even necessary, to apply radiation at a second frequency, or range of frequencies. The term homonuclear double resonance refers to experiments in which the two forms of irradiation are applied to the same nuclear species (e.g. to protons); otherwise, the term heteronuclear double resonance is used. Here, we briefly discuss some of the more widespread applications of such experiments.

(i) It was mentioned in § 6.2.4 that spin decoupling can be of great value as a means of assigning and simplifying spectra. Perhaps the most common usage of spin decoupling is in ^{13}C n.m.r. studies, where ^{1}H–^{13}C couplings are removed by irradiation of the appropriate ^{1}H resonance frequencies. We can selectively irradiate specific ^{1}H resonances, or alternatively remove all ^{1}H–^{13}C couplings by irradiating the whole of the ^{1}H spectral range. In such experiments, the ^{1}H irradiation is generally modulated by noise, and it is the frequency distribution of this noise modulation that determines the effective bandwidth of the ^{1}H irradiation. Decoupling experiments require that the irradiation be applied during the actual free induction decay; this can lead to problems (e.g. with leakage, see § 7.3.1) if one wishes to do homonuclear decoupling, but these problems are by no means insuperable.

(ii) As mentioned in § 6.6, double irradiation can also lead to enhancements in signal intensities via the nuclear Overhauser effect. Thus for example, continuous irradiation of ^{1}H signals in a ^{13}C n.m.r. experiment will cause decoupling of the ^{13}C signals, and will also generate an Overhauser effect. Sometimes, however, we may wish to spin decouple without obtaining an Overhauser effect, while in other experiments the reverse may be desirable. The required effects can be achieved by gated irradiation as illustrated in Fig. 7.9.

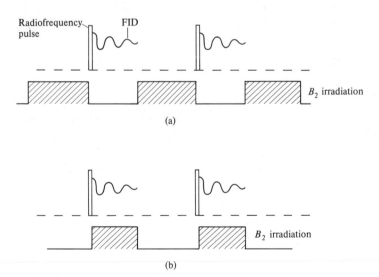

Fig. 7.9. Diagram illustrating two different types of double resonance experiment. (a) Gated irradiation producing an Overhauser effect without spin decoupling. This type of gating is also used for saturation transfer experiments and for solvent suppression. (b) Gated irradiation producing spin decoupling without any Overhauser effect.

(iii) In saturation transfer experiments, selective irradiation is applied to one signal, and the effect produced on a second signal can be interpreted in terms of chemical exchange between the two species that generate these signals (see Appendix 6.1). The gating procedure that is employed is similar to that shown in Fig. 7.9(a). Similar information can be achieved with the use of selective or non-selective 180° pulses (see Campbell, Dobson, Ratcliffe, and Williams 1978).

(iv) A large solvent peak (e.g. the ^{1}H resonance of H_2O) can be significantly reduced in intensity by applying selective saturating irradiation to the signal. Again, the irradiation must be gated on in a similar manner to that shown in Fig. 7.9(a). (The ^{1}H signal of water can also be reduced by means of the spin-echo technique, which has been used particularly effectively in studies of cellular suspensions at high frequencies (Rabenstein and Isab 1979; see also Fig. 3.6). Alternatively, the so-called 'Redfield 214 pulse' can be used; this is a radiofrequency pulse, the shape and frequency of which are chosen in such a way that the water signal is not excited (Redfield, Kunz, and Ralph 1975). Finally, the problem can be circumvented by using correlation spectroscopy (see § 5.6).)

Many of these double resonance experiments require the use of frequency-selective irradiation. It should be remembered that the selectivity of a radiofrequency pulse depends upon its duration; the longer the pulse, the more selective it can be (see § 5.5.1).

7.3. The receiver

7.3.1. The basic design

A simplified block diagram of the receiver is shown in Fig. 7.10. This diagram is not truly representative of all n.m.r. receivers, but it does serve to illustrate most of the important principles of signal reception.

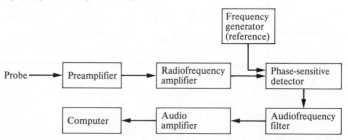

Fig. 7.10. A simplified block diagram of the spectrometer receiver.

The signal and noise emerge from the probe in the form of a very small voltage, typically a few nanovolts. It is therefore essential to ensure that in the early stages of amplification very little additional noise is introduced because only a small noise level superimposed upon a few nanovolts could greatly degrade the signal-to-noise ratio that is finally obtained. In this respect, the preamplifier in particular must be designed and built with great care. The function of the preamplifier and the radiofrequency amplifier is therefore to amplify the signal and noise without having any adverse effect on the signal-to-noise ratio; this ratio should be determined entirely by the characteristics of sample and probe.

At this stage it is appropriate to consider the nature of the spectrum we hope to obtain. Typically, we might expect to observe a spectrum at, say, 100 MHz which contains a number of resonances of width a few hertz occupying a total spectral width of a few kilohertz. However, the computer is not capable of detecting and distinguishing between signals at such high frequencies, nor is the filter capable of selecting a region occupying only a few kilohertz at a frequency of 100 MHz. Therefore the approach taken is to subtract the frequency ν_1 of the applied B_1 field from the frequency ν_0 of the signals, thereby generating a group of frequencies in the range zero to a few kilohertz. These frequencies, which are in the audiofrequency range, can be handled by the computer and can be filtered adequately. The required subtraction of frequencies is accomplished with the aid of a device known as a phase-sensitive detector. This device multiplies the signal alternately by ± 1 with a frequency of alternation known as the reference frequency which in the simplest case is equal to the frequency ν_1 of the field B_1. Mathematical analysis shows that there is an output from the phase-sensitive detector of the form $\cos\{2\pi(\nu_0 - \nu_1)t + \phi\}$. This represents a signal of frequency $\nu_0 - \nu_1$ which is equal to the difference in frequency between the signal and the reference. The phase angle ϕ is equal to the phase difference between signal and reference, hence the name of the device.

The audiofrequency output from the phase-sensitive detector is filtered and then amplified to the required level (usually about 2 V peak-to-peak) before being fed into the computer. The filter removes high-frequency noise, which if present would have adverse effects on the signal-to-noise ratio (see § 7.4.1).

Often, there will be a constant d.c. level observed at the audiofrequency stage of the receiver. This can occur as a result of unintended 'leakage' of the reference frequency into the signal channel of the receiver. Phase-sensitive detection of this leakage generates a signal of zero frequency, resulting in a d.c. output. This undesired effect can be minimised by adjusting the d.c. level to zero prior to starting an accumulation. Pulse sequences that are now routinely used will greatly reduce the effect, so that it should not be possible to detect the effects of leakage in the final spectrum. However, if a signal does appear in any spectrum at zero frequency, care should be taken to ensure that it is a genuine signal and not simply an effect of leakage.

To further reduce the leakage and for additional technical reasons, many spectrometers convert from radiofrequencies to audiofrequencies in two stages; the radiofrequency signal is first converted to an intermediate frequency, such as 1 MHz, before subsequent conversion to the audiofrequency signal.

7.3.2. *The phase-sensitive detector and the use of quadrature detection*

An important drawback of the detection system considered above is that information is lost as to whether $\nu_0 - \nu_1$ is positive or negative. Therefore, following phase-sensitive detection and Fourier transformation, all signals will appear to one side of the reference frequency, regardless of the sign of $\nu_0 - \nu_1$. This effect is illus-

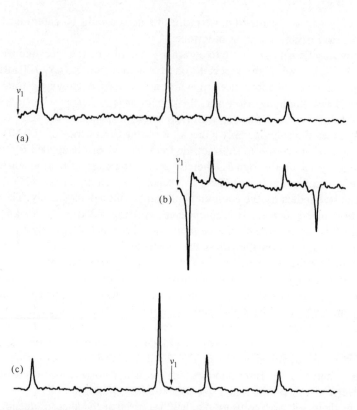

Fig. 7.11. The positioning of the reference frequency. (a) The reference is to one side of the signals, and the signals are displayed correctly. (b) The reference is positioned among the signals, and this causes problems (when not using quadrature detection; see § 7.3.3). (c) When using quadrature detection the reference can be placed among the signals.

trated in Fig. 7.11. Figure 7.11(a) shows a spectrum containing four signals detected in the manner described above with the reference frequency positioned to one side of the lines. Figure 7.11(b) shows the effect of placing the reference among the signals. Two of the signals are positioned correctly but the positions of the other two have been reflected about the reference frequency, an effect known as folding- over. These signals are also incorrectly phased. Placing of the reference frequency anywhere in the middle of a spectrum therefore results in incorrect positioning and phasing of some of the spectral lines and so it is imperative when using this method of detection to ensure that the reference frequency (which, to recapitulate, is the frequency of the applied B_1 field) is positioned to one side of the whole spectrum.

It is also important to appreciate that, for exactly the same reasons, noise is reflected about the reference frequency. Therefore two sets of noise contribute to the spectrum instead of one. Since noise is random, this causes the noise amplitude to increase by a factor of $\sqrt{2}$ and the signal-to-nose ratio to decrease by a factor of $\sqrt{2}$.

Clearly, it would be advantageous to employ a method that is capable of over-coming this problem. The use of two phase-sensitive detectors in quadrature provides such a method. Signal and reference are fed into two separate phase-sensitive detectors, and as illustrated in Fig. 7.12 the phase of the reference is

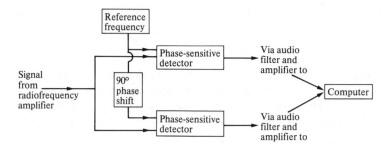

Fig. 7.12. Simplified block diagram for a receiver employing quadrature detection.

shifted by $90°$ in one of the detectors. It is this $90°$ phase shift to which the term 'in quadrature' refers. The outputs from the two phase-sensitive detectors differ in phase by $90°$, but should be similar in all other respects. They are both filtered and amplified before being fed into two separate channels of the computer.

Figure 7.13 shows FIDs observed in the two channels (a) when $\nu_0 - \nu_1 > 0$ and (b) when $\nu_0 - \nu_1 < 0$. The signals in Channel 1 are the same in the two cases but those in Channel 2 are not; when $\nu_0 - \nu_1 > 0$ the phase of the signal in Channel 2 leads that in Channel 1 by $90°$, whereas when $\nu_0 - \nu_1 < 0$ the phase in Channel 2 lags that of Channel 1 by $90°$. It is this difference that, on Fourier transformation, leads to the separation of positive and negative frequencies.

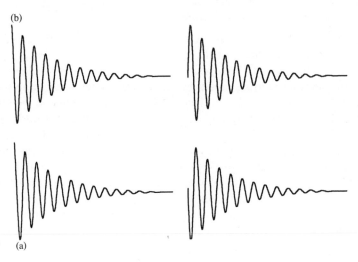

Fig. 7.13. Free induction decays observed in the two channels when employing quadrature detection. (a) $(\nu_0 - \nu_1) > 0$, and (b) $(\nu_0 - \nu_1) < 0$.

The effects of this method of separation are illustrated by the spectrum of Fig. 7.11(c) which was obtained under similar conditions to those used for the spectrum of Fig. 7.11(b), except that quadrature detection was used. In contrast to the spectrum of Fig. 7.11(b), no folding-over is observed when the reference frequency is in the middle of the spectrum. In fact spectra obtained with quadrature detection have zero offset at the centre; negative frequencies are displayed to the right and positive frequencies to the left.

Quadrature detection has two main advantages over detection using a single phase-sensitive detector. Firstly, noise is not folded about the reference frequency, and as a result there is an improvement in signal-to-noise ratio of a factor of $\sqrt{2}$ for a given number of scans. As can be seen from § 1.7.2, this corresponds to a saving in accumulation time of a factor of 2, which can be extremely valuable. Secondly, the reference frequency can be placed in the middle of the spectrum. As a result, the maximum offset in a spectrum of width W kHz is only $W/2$ kHz, as opposed to W kHz when a single-phase sensitive detector is used. This considerably reduces bandwidth problems that may arise from the effects of finite pulse width (see § 7.2.5). The ability to place the reference frequency anywhere in a spectrum can also be useful for double irradiation experiments.

A potential difficulty with the use of quadrature detection is that unless the two signals in the computer have exactly equal amplitudes and are exactly 90° out of phase with each other, one will obtain images of the signals reflected about zero frequency. For this reason it is necessary to employ special data routing techniques to correct for inaccuracies in the gains and phase settings of the receiver channels (Hoult and Richards 1975; Stejskal and Schaefer 1974). Quadrature detection in conjunction with such data routing techniques is now a routine feature of modern n.m.r. spectrometers.

7.3.3. Adjustment of the receiver settings

The variables that generally require adjustment are the gain (amplification) settings at the radiofrequency, intermediate frequency, and audiofrequency stages, the filters, and the d.c. level. As discussed in § 7.3.1, the d.c. level should be adjusted to zero, but slight mis-setting should be unimportant when using pulse sequences that reduce the effects of leakage. The role of the filters is related to the manner in which the data is collected by the computer, and a discussion of filter settings is therefore given in § 7.4.1. The gain settings should be low enough to ensure that the signal or noise does not overload any of the receiver stages. At the same time, the settings should be sufficiently high to ensure that the receiver itself contributes a negligible amount of noise. In general, therefore, we should aim to achieve conditions where each of the stages is not quite overloaded. A further constraint is that the computer will be unable to cope with signals above a given voltage (often the limits will be ±2 V), and the audiofrequency gain should be set accordingly.

7.4. The computer

The computer of a modern n.m.r. spectrometer has a wide range of functions. Its main functions are (i) to control the timing of the experiment, for example to control the pulse width and interval, (ii) to accumulate the data, and (iii) to process and display the data. In § 7.2 we discussed how to optimize the pulse width and interval; here we discuss the latter two functions of the computer. Further details of n.m.r. computers are given by Cooper (1976).

7.4.1. Collection of data

The signals that emerge from the receiver are continuous whereas the computer samples the incoming signals at discrete time intervals, and it is particularly important to appreciate the implications of this.

Consider a FID from a single resonance, as shown in Fig. 7.14(a). The interval between successive samplings of the decay is set by the operator and is commonly known as the dwell time. It is important to know how frequent the sampling must be in order for the resulting discrete pattern to represent correctly the signal that is

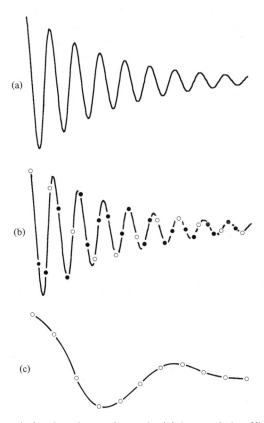

Fig. 7.14. The free induction decay shown in (a) is sampled sufficiently rapidly in (b) to ensure that the decay is interpreted correctly. The sampling points in (c) are identical to the 1st, 4th, 7th, 10th, etc. points of (b) (open circles); i.e. the sampling frequency in (c) is one-third of the sampling frequency in (b). Clearly, with this slower sampling frequency the free induction decay is interpreted incorrectly, for its apparent frequency is too low.

being sampled. There is a theorem due to Nyquist which shows that the sampling must be performed at least twice per wave cycle in order for the resulting frequency to be equal to the actual frequency of the wave. This is illustrated in Fig. 7.14. In Fig. 7.14(b) the sampling of the FID from above is sufficiently rapid, whereas in Fig. 7.14(c) it is not; the apparent frequency in the latter case is less than it should be because the frequency of sampling is less than twice the actual frequency of the waveform. Thus the sampling interval, or dwell time, determines the maximum frequency observed after computer detection; if the dwell time is d s, the maximum frequency is $1/2d$ Hz. Therefore, if a single-phase sensitive detector is used in the receiver the final spectral width, commonly called the sweep width, is $1/2d$ Hz. If quadrature detection is employed the spectral width is twice this value, i.e. $1/d$ Hz, because both positive and negative frequencies are displayed (see § 7.3.1). The implications of this are illustrated by the spectra shown in Fig. 7.15. Figure 7.15(a) shows a spectrum obtained using a dwell time of $330\,\mu$s, corresponding to a sweep width of 3 kHz. The signals are all positioned correctly in the spectrum. Figure 7.15 (b) shows a spectrum obtained under identical conditions, except that a dwell time of $500\,\mu$s was used. This is clearly too long for this particular spectrum, for it can be seen that the frequency of signal A is interpreted incorrectly by the sampling process. This effect, known as aliasing, illustrates the importance of setting the dwell time correctly.

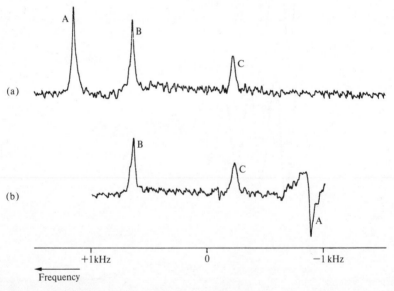

Fig. 7.15. (a) shows a spectrum obtained using a dwell time of $330\,\mu$s, corresponding to a spectral width of 3 kHz. The signals are all positioned correctly. (b) shows a spectrum obtained under identical conditions, except that a dwell time of $500\,\mu$s was used. Signal A should be at $+1125$ Hz. In (b), the sampling effect causes the signal to be reflected about the end of the spectrum, i.e. about the frequency of 1 kHz, producing what might be expected to be an apparent value of 875 Hz. However, there is an additional effect when using quadrature detection that causes the signal to appear, not at 875 Hz, but at -875 Hz.

Aliasing has additional important implications, for in addition to affecting signals, it also affects noise. For example, if we select a sweep width of ±2.5 kHz, the noise at frequencies beyond ±2.5 kHz will be aliased back into the spectrum to appear as relatively low frequency noise. It is therefore essential that high-frequency noise be filtered out before entering the computer, otherwise the effect of aliasing would be to increase the noise level significantly and hence to decrease the final signal-to-noise ratio. The receiver filters should therefore be set to accept all frequencies within the sweep width but to reject all others. The effects of incorrect filter setting are illustrated in Fig. 7.16. Figure 7.16(a) shows a spectrum obtained under correct conditions; the sweep width is 5 kHz and the filters are set to the same frequency. Figure 7.16(b) shows a spectrum obtained under identical conditions, except that the filters were set to 10 kHz. The resulting aliasing of noise causes the observed increase in the noise level. Figure 7.16(c) shows what happens when the filters are set to 3 kHz; there is a large reduction in intensity of both signal and noise at the extremities of the spectrum and these signal intensities no longer reflect the true intensities that should be obtained.

Although the filters are designed to have a very sharp cut-off frequency, they may unavoidably reduce the intensity of signals that are positioned close to the cut-off frequency. For this reason, if accurate intensity measurements are required, it is safest to select a sweep width and filter setting that are at least 25 per cent greater than the frequency distribution of the spectral lines.

Another variable that must be set is the number of sampling points. A FID should be collected for at least $4\,T_2^*$ to ensure that essentially the whole decay is collected. Thus, if T_2^* is 50 ms, the acquisition time should be about 200 ms. For a spectrum of width 5 kHz, the dwell time is 200 μs if quadrature detection is used.

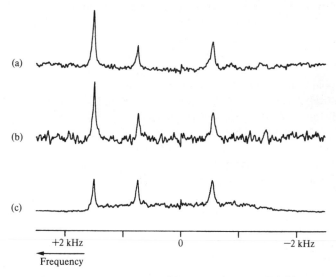

Fig. 7.16. The effects of filter settings: (a) correct setting; (b) filters set too high a frequency bandwidth result in additonal noise; and (c) filters set too low a bandwidth reduce signal and noise at the extremities of the spectrum.

Therefore, 1000 sampling points per channel would be required for a total acquisition time of 200 ms (but it may be preferable to use more points and a longer acquisition time; see § 7.4.2). If T_2^* were longer, than more points, e.g. 2000–4000, would be necessary. In fact, a computer operates in terms of powers of 2, and the appropriate number of points would be 1024, 2048, and 4096, corresponding to 2^{10}, 2^{11}, and 2^{12}. These are generally termed $1K$, $2K$, and $4K$ points respectively.

One additional variable is the delay time which represents the interval between the end of the radiofrequency pulse and the beginning of acquisition of data. A delay is often required to ensure that the probe and receiver have recovered from the direct effects of the high-voltage radiofrequency pulses which are experienced regardless of whether or not a sample is present. Often, the delay time is set to be approximately equal to the dwell time. Its setting affects the phase of the spectral lines, as we shall see later in this chapter.

7.4.2. Data processing

Following accumulation of the FID, the resulting signal must be processed in order to produce a spectrum in which signal amplitude is plotted as a function of frequency. In addition to the process of Fourier transformation, a number of additional steps are usually required, and in this section we discuss briefly the important aspects of data processing.

Fourier transformation

Computers designed for use with n.m.r. spectrometers have a built-in Fourier transformation programme, which employs an algorithm devised by Cooley and Tukey (1965), which is known as the fast Fourier transform. The Fourier transform always contains the same number of data points as were transformed. For example, transformation of two decays collected in quadrature, each containing $2K$ points, produces real and imaginary parts each containing $2K$ points (see § 5.5.2 for a discussion of these real and imaginary components). It can readily be shown from this and from the discussion of § 7.4.1 that the frequency separation between adjacent points in the transformed spectrum is equal to 1/acquisition time. If the acquisition time were equal to $4T_2^*$ for a given signal, this would imply that the frequency separation of the points in the transform would be similar in magnitude to the width of the signal. Thus the signal would contain very few data points, and would be rather poorly defined. For this reason, it is preferable to increase the acquisition time by increasing the number of sampling points in the free induction decay; alternatively, one can zero fill (see below).

Prior to Fourier transformation, it is advisable to store each spectrum on magnetic disc or tape units if these are available. This is because data processing, together with any errors that may occur in the course of processing, can lead to irreversible loss of information.

Zero filling

A procedure known as zero filling is sometimes carried out prior to Fourier transformation. This process invoves adding an array of n zeros to the end of each FID of n data points. This has the effect of doubling the number of points in the transformed spectrum and can lead to an improvement in the quality of the spectrum (Bertholdi and Ernst 1973). However, if substantial exponential filtering is performed (see below), zero filling will simply generate additional interpolated points in the transformed spectrum and will provide little, if any, further information.

Phase correction

The aim of phase correction, as mentioned in § 5.5.2, is to ensure that all the signals in the real part of the transformed spectrum are in the absorption mode. We saw that the phase correction is necessary because a typical signal will appear as in Fig. 5.13, rather than as in Fig. 5.11. The difference in phase between these two signals represents the phase correction that must be applied to the transformed spectrum.

Now let us consider the effect of a finite delay time between the end of the radiofrequency pulse and the start of data acquisition. If there is a delay time t, then in this time a waveform of frequency ν goes through νt cycles and suffers a phase shift of $2\pi\nu t$ rad. A phase correction of this amount is therefore introduced as a result of this delay. It should be noted that resonances of different frequencies suffer different phase shifts; in fact the delay introduces a phase shift that is directly proportional to frequency. This can be corrected for by the so-called frequency-dependent phase correction. The use of a constant phase correction, together with a phase correction that is linearly proportional to frequency, should ensure that all resonances within a spectrum are correctly phased.

The phase correction makes use of both real and imaginary parts of a spectrum. A phase change of ϕ corresponds to manipulation of each data point according to the following equations:

$$r_2 = -i_1 \sin\phi + r_1 \cos\phi$$

$$i_2 = r_1 \sin\phi + i_1 \cos\phi$$

where r_2 and i_2 correspond to the real and imaginary parts of each point of the spectrum following the phase correction and r_1 and i_1 correspond to the points prior to correction. It is therefore essential to preserve both real and imaginary components of the spectrum until phase correction is complete.

Manipulation of the FID

The quality of n.m.r. spectra is ultimately determined by the signal and noise that are stored within the computer. However, certain characteristics of the spectra, such as the signal-to-noise ratio or the resolution, can be considerably improved by means of appropriate manipulation of the data prior to Fourier transformation. In this section we illustrate the basic principles behind some of the methods of data manipulation.

(i) Sensitivity enhancement

Consider a simple FID and its transform as shown in Fig. 7.17(a). Suppose we wish to improve the signal-to-noise ratio of the spectrum. This can be achieved by multiplying the FID point by point by a decaying exponential as shown in Fig. 7.17 (b). The function of this multiplication is to eliminate the noise preferentially by lending more weight to the initial part of the FID, where the signal-to-noise ratio is

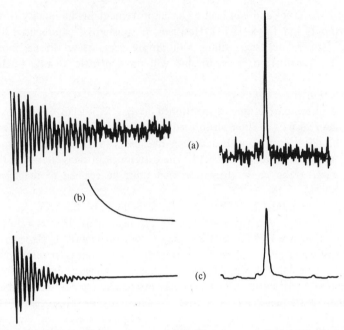

Fig. 7.17. (a) A free induction decay and its Fourier transform. The free induction decay in (c) is obtained on multiplying the decay of (a) point by point by the exponential decay of (b). This results in a signal of enhanced signal-to-noise, but also of enhanced linewidth.

high, than to the latter part, where the signal-to-noise ratio is much lower. The effects of exponential multiplication are shown in Fig. 7.17(c). There are two important features to note when comparing the transformed spectrum with that shown in Fig. 7.17(a). Firstly, the signal-to-noise ratio is considerably enhanced by performing the multiplication. Secondly, however, the linewidth is increased as a result of this process. The reason for this is that multiplication by the decaying exponential has the effect of reducing the time constant of the FID. As we have seen, there is a reciprocal relationship between the time constant of the decay and the linewidth, and therefore the linewidth increases correspondingly. In fact, the exponential multiplication increases the linewidth by an amount $\Delta \nu = 1/\pi T$, where T is the time constant of the exponential decay.

It can be seen from Fig. 7.17 that exponential multiplication in the time domain has the same effect as filtering would have in the frequency domain, and in fact the process is sometimes referred to as filtering. One of the virtues of Fourier transform n.m.r. is the ease with which this process is achieved, for filtering in the frequency domain is rather more complicated. (It should be noted that this filtering process is completely different from the function of the filters used in the spectrometer receiver; these latter filters are discussed in § 7.4.1.)

For an optimal signal-to-noise ratio in the transformed spectrum, the decaying exponential should have the same time constant as the FID. The effect of this so-called 'matched filter' is to double the linewidth, and therefore the resolution is significantly degraded by the process. If a spectrum contains several resonances of differing linewidths, it is clearly impossible to optimize the time constant of the exponential multiplication for all of the lines. Under these circumstances we could process the spectrum several times using different time constants in each case. Usually, however, this should be unnecessary as the signal-to-noise ratio achieved in the final spectrum is not very sensitive to mis-setting of the time constant of the multiplication.

(ii) Resolution enhancement

Figure 7.17 illustrates the important rule that enhancement of the signal-to-noise ratio can only be achieved at the expense of resolution. The reverse of this statement also holds, as is readily demonstrated by all of the methods that are available for resolution enhancement. Let us now consider how resolution enhancement can be achieved.

The simplest method is to multiply the FID by a growing exponential rather than by the decaying exponential of Fig. 7.17(b). However, this method can lead to problems; in particular, there will be distortion of the Fourier transformed spectrum

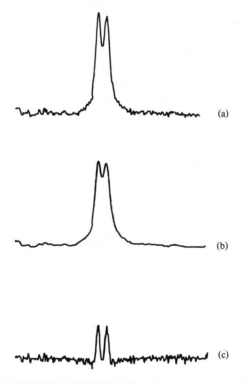

Fig. 7.18. Resolution enhancement by the convolution difference technique (Campbell, Dobson, Williams, and Xavier, 1973). One copy of the free induction decay is multipled by an exponential decay of time constant T_A, and on Fourier transformation yields spectrum (a). If a shorter time constant T_B is used on the duplicate decay, spectrum (b) is obtained. Subtraction of (b) from (a) gives (c). (c) clearly has better spectral resolution than (a).

if the signal is non-zero at the end of the period of data acquisition. For this and other reasons, a variety of alternative techniques for resolution enhancement have been devised (Campbell *et al.* 1973; DeMarco and Wuthrich 1976; Ernst 1966; Ferrige and Lindon 1978). One of these methods, termed convolution difference, is illustrated in Fig. 7.18. The FID is duplicated within the computer, and one copy is multiplied by an exponential decay of time constant T_A and then Fourier transformed to give spectrum (a). The duplicate FID is then multiplied by a second decaying exponential of time constant T_B and is also transformed, giving spectrum (b). The second spectrum is multiplied by a constant K, which may or may not be equal to unity, and is then subtracted from the first. The nature of the resulting spectrum (Fig. 7.18(c)), is dependent on the magnitude of T_A, T_B, and K, but again significant resolution enhancement is obtained only at the expense of a large reduction in the signal-to-noise ratio. The technique is highly versatile and is routinely used not only for enhancing resolution but also for eliminating particularly broad components of spectra. However, the method necessarily generates some distortion of the spectral lines, and great care must be taken when interpreting the resulting signal intensities. Indeed, if $K=1$ the total integrated intensity of any signal is zero because the wings of each resonance are negative.

7.5 Summary of spectrometer operation

We conclude this chapter by listing the various steps that should be taken when performing a routine n.m.r. experiment on a typical spectrometer. We assume that the sample has been set up correctly within the probe, and that the probe has been tuned and matched at the required frequency (see Chapter 8).

Setting up the spectrometer

(i) Lock the spectrometer, if necessary, using the field-frequency lock.

(ii) Optimize the homogeneity of the field B_0, using the shim coils.

(iii) Select a suitable frequency offset for the radiofrequency pulses.

(iv) Select the dwell time (which defines the sweep width), delay time, filter settings, and receiver amplification.

(v) For double resonance experiments, select the frequency, timing, bandwidth, and power of the second irradiation frequency.

(vi) If unknown, determine the length of the $90°$ pulse.

(vii) Select the pulse width and interval, and the number of pulses to be applied.

(viii) Collect data.

Data processing

(i) Store the accumulated free induction decay on magnetic disc or tape.

(ii) Manipulate the free induction decay as required, e.g. by applying an exponential multiplication.

(iii) If desired, zero fill.

(iv) Fourier transform the data.

(v) Perform the phase correction.

(vi) Display and plot the spectrum as required.

8

Probe design

Probe design deserves a chapter to itself because of its supreme importance in studies of living systems. Improvements in probe design can greatly enhance the versatility and performance of a spectrometer, and one of the main purposes of this chapter is to convince the reader of the surprising ease with which these improvements can be made.

In many ways the probe represents an interface between magnet, transmitter, and receiver. It is usually shaped in the form of a cylinder that fits snugly inside the bore of the magnet and is made of a non-magnetic metal such as aluminium. As mentioned briefly in § 1.7, a key feature of the probe is a coil of wire that either surrounds the sample or is positioned adjacent to it (see Fig. 8.1). This coil transmits radiofrequency power into the sample in the form of the radiofrequency field

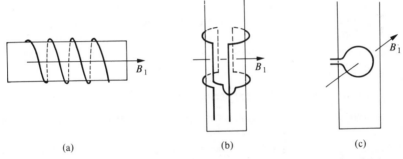

(a)   (b)   (c)

Fig. 8.1. Radiofrequency coil designs. The usual types are (a) solenoidal, or (b) saddle-shaped. Both of these coils are designed to surround a cylindrical sample. (c) A more unusual type of coil, termed a surface coil, consists of a loop of wire, of one or more turns, that is placed adjacent to the sample.

B_1 and detects the resulting signal. The probe also contains a small amount of electronic circuitry which associated with the coil makes up the tuned circuit(s) that are required for efficient signal detection. In addition, it may contain suitable devices for adjusting and controlling the temperature and for spinning the sample, together with any perfusion tubing, leads for electrical stimulation, etc. that may be required for studies of living systems.

8.1. Why bother with probe design?

This is a most pertinent question in view of the quality of the commercial probes that are readily available. There is no doubt that, for *solution* studies, the commercial probes that are routinely used in most spectrometers provide spectra of excellent quality. Any lack of versatility that they may have in terms of sample volume or shape is relatively unimportant because of the adaptability of the samples. However, for studies of *living systems* the situation is very different. For example, a perfused heart may not fit inside the standard n.m.r. tubes that are used for ^{31}P or ^{13}C n.m.r. studies. A liver simply has the wrong shape for the cylindrical n.m.r. tubes that must be used in commercially available probes. An intact rat certainly does not fit inside a conventional probe! Furthermore, many studies of living systems require perfusion and electrical leads, and it may be difficult to accommodate these within the probes that are available.

In general, therefore, the necessity of finding a sample to suit the probe severely limits the choice of preparation. Moreover, it is likely that those experiments that are feasible will be performed under conditions that are far from ideal, both in terms of the physiology and the quality of the spectra. If studies of living systems are to be done optimally, it is essential to be able to design the probe to suit the physiological conditions rather than vice versa. This is the main incentive behind designing new probes. However, it is also clear that if a new probe is to be made due consideration should be given to designing it in such a way that the quality of the spectra is optimized.

There is no doubt that the quality of n.m.r. studies of living systems has been revolutionized by this type of approach, and that the effort involved in constructing new probes is totally justified by the enhanced versatility and quality of the experiments. In the following sections we discuss some of the important principles governing probe design and describe examples of probes that have so far proved highly successful.

8.2. Considerations of spectrometer performance

The spectrometer performance characteristics that must be considered when designing a probe are as follows.

(i) Sensitivity: it is essential to optimize signal-to-noise ratios if we wish to obtain acceptable spectra in a minimum time.

(ii) B_0 homogeneity: inhomogeneity in the static field B_0 leads to broadening of the spectral lines as mentioned in § 7.1. It is important to ensure that this broadening mechanism makes only a small contribution to the observed linewidths, otherwise the resolution and the signal-to-noise ratios will be significantly reduced.

(iii) B_1 homogeneity: if the B_1 field is not uniform throughout the sample volume, different parts of the sample will contribute unequally to the observed signal. In addition to complicating measurements of concentrations, this can lead to other problems. For example, it may be difficult to make accurate measurements of relaxation times (see §§ 6.3.6 and 6.3.7).

All of these important characteristics are greatly affected by probe design, but unfortunately a compromise between them is often necessary; a probe designed to optimize sensitivity will not necessarily produce optimal B_0 homogeneity. It is therefore essential to consider the specific requirements of each experiment when designing probes. For example, ^{31}P n.m.r. studies of metabolism in living systems require excellent sensitivity, but a relatively modest B_0 homogeneity of about 1 part in 10^7 is usually sufficient. For these experiments a radiofrequency coil made of thick (about 3 mm in diameter) copper wire should be suitable (the thicker the wire the better the sensitivity of the coil (see Appendix 8.1)). However, with this thick wire it is extremely unlikely that the type of resolution required for high-resolution ¹H n.m.r. studies (a few parts in 10^9) will be achieved, and for such studies much thinner wire, or preferably foil, should therefore be employed.

In order to design the most suitable radiofrequency coils, it is useful, albeit not essential, to have an understanding of the theory of signal reception. The approach presented here is that developed by Hoult and Richards (1976). Their most elegant and informative paper is thoroughly recommended to anyone interested in probe design and the more general problem of sensitivity.

8.3. The signal-to-noise ratio

Let us consider a solenoidal coil surrounding a sample as shown in Fig. 8.1. Suppose that a 90° pulse has been applied by passage of a suitable current through the coil. This coil is also used for detecting the resulting magnetization that is generated in the xy-plane. The magnetization precesses with angular frequency ω_0 and induces an electromotive force (e.m.f.) of the same frequency in the coil. The magnitude ϵ of this e.m.f. is given by Faraday's law of electromagnetic induction:

$$\epsilon = -d\phi/dt. \tag{8.1}$$

where ϕ is the magnetic flux linkage through the coil. The e.m.f. ϵ constitutes the signal that, after much processing, appears in the final spectrum. Now it is perfectly reasonable to expect that the magnetic flux passing through the coil, and hence the induced e.m.f., would decrease if the coil were moved away from the sample; we could say that under these conditions the coupling between sample and coil would decrease. Similarly, as the coupling decreases the B_1 field that would be produced at the sample if *unit* current were passed through the coil would also decrease. It turns out that there is a well-defined relationship between the magnitude of the induced e.m.f. and the strength of the B_1 field generated at the sample by unit current in the coil. This relationship was used by Hoult and Richards (1976) to obtain an expression for the induced e.m.f. The details of their calculation will not be presented here; instead, we simply quote the result for a sample of volume V_s situated in a region of homogeneous B_1 field. Under such conditions the e.m.f. induced in the coil immediately after application of a 90° pulse is

$$\epsilon = \omega_0 B_{1(u)} M_0 V_s \tag{8.2}$$

where $B_{1(u)}$ is the field generated in the xy-plane by unit current passing through the coil and V_s is the sample volume that experiences this field. The nuclear magnetization M_0 is given by

$$M_0 = C\gamma^2\hbar^2 I(I+1)B_0/3kT_s. \tag{8.3}$$

where C is the number of resonant nuclei per unit volume, γ is the magnetogyric ratio of these nuclei, T_s is the sample temperature, and k is the Boltzmann constant.

Equations (8.2) and (8.3) enable the signal generated in the coil to be evaluated in terms of parameters that are either known or can be estimated. In order to evaluate the signal-to-noise ratio we now have to obtain an expression for the noise that is generated in the experiment. In a correctly designed spectrometer almost all of the noise should originate in the radiofrequency coil itself (ignoring for now the noise generated by the sample which is discussed in § 8.5). Therefore the problem reduces to one of estimating the noise developed in the coil. This noise consists of random fluctuations of e.m.f. associated with the Brownian motion of the electrons within the coil. It can be shown that the root mean square noise e.m.f. over a bandwidth $\Delta \nu$ at the resonant frequency is given by

$$N = (4kT_c R_c \Delta \nu)^{1/2} \tag{8.4}$$

where R_c is the resistance of the coil and T_c is its temperature.

If we wish to optimize the signal-to-noise ratio available from a single radio-frequency pulse, it is therefore necessary to maximize the ratio ϵ/N. At a specified field strength, frequency, and temperature this reduces to maximizing the ratio $CV_s B_{1(u)}/R_c^{1/2}$; all of the other parameters are independent of the sample and of the coil design.

As we shall see, the beauty of this approach to the problem of sensitivity is that all of the parameters in the above expressions are either known or can be measured or calculated with a fair degree of accuracy; no guesswork is required. Simple experiments can be performed (see § 8.7.1), without any requirement for a spectrometer, to make an accurate assessment of the relative merits of different probe designs.

Before discussing the various types of coil design, let us consider the nature of the ratio $CV_s B_{1(u)}/R_c^{1/2}$. The first two parameters C and V_s represent the concentration of resonant nuclei and the sample volume, and their product is simply the total number of resonant nuclei within the sample. The other two terms $B_{1(u)}$ and R_c are characteristic of the coil. At first sight, this expression suggests that the signal-to-noise ratio will increase linearly with sample concentration and volume. However, this volume dependence does not in fact hold, the reason being that $B_{1(u)}$, R_c, and V_s are not independent variables. Firstly $B_{1(u)}$ and R_c are dependent upon the dimensions of the coil, and hence in general upon the sample volume. Moreover, as we shall see in § 8.5, the sample itself can also contribute towards the effective resistance and this contribution is dependent upon $B_{1(u)}$ and V_s. Therefore it is not intuitively obvious from this equation to what extent the signal-to-noise ratio will increase with the sample volume. In practice, however, we can expect the signal-to-noise ratio to increase very approximately according to $V_s^{1/2}$, but the dependence may be stronger for very small samples and very much weaker for large conducting samples (see § 8.5.1).

8.4. The various types of coil design

8.4.1. Solenoidal and saddle-shaped coils

Conventional radiofrequency coils are usually either solenoidal or saddle-shaped (see Fig. 8.1). The saddle-shaped coil is sometimes rather loosely referred to as a Helmholtz coil.

Almost all superconducting magnets have their field B_0 directed vertically, and therefore the radiofrequency field B_1 must be applied in the horizontal plane. A saddle-shaped coil can be oriented as shown in Fig. 8.1(b), which is convenient as the sample tube can be positioned vertically within the probe. However, when a solenoidal coil is used the sample tube must be mounted horizontally, and this has a number of disadvantages. Firstly, the cylindrical symmetry of the probe is destroyed, and this could adversely affect the homogeneity of the field B_0 resulting in loss of spectral resolution. Secondly, although sample spinning is possible (see Oldfield and Meadows 1978), it is rather more difficult to achieve than when the sample tube is held vertically. Thirdly, the diameter of the probe, and hence the bore of the magnet, must be fairly large in order to accommodate a solenoid and horizontal tube. Finally, there may be problems associated with the physiological techniques employed; for example, it is far more difficult to measure muscle tension when the muscles are mounted horizontally than when they are positioned vertically. At the time of writing commercial probes used with superconducting magnets routinely employ saddle-shaped rather than solenoidal coils because of the relative simplicity of using the former design.

The great advantage of solenoidal coils is that they produce greater signal-to-noise ratios than saddle-shaped coils for a given sample volume. This has been shown theoretically and experimentally by Hoult and Richards (1976), and has subsequently been confirmed in a number of laboratories. The reason for this is that the ratio $B_{1(u)}/R_c^{1/2}$, which reflects the signal-to-noise characteristics of a coil (see § 8.3), is greater for a solenoidal than for a saddle-shaped coil; essentially, a saddle-shaped coil requires a longer length of wire to generate a given $B_{1(u)}$ field and hence has greater resistance than a solenoidal coil. Hoult and Richards (1976) predicted a difference in the signal-to-noise ratio by a factor of up to 3, and in general it appears that, for optimally designed coils, the sensitivity of a solenoid can be expected to be better by a factor of about 1.5–2. This corresponds to a reduction in accumulation time by a factor of about 2–4, which is of tremendous value for a wide range of studies. It is for this reason that solenoids have recently been used to great advantage. Of course, for magnets in which the field B_0 is horizontal a vertical solenoidal coil can be, and is, used.

Therefore, when sensitivity is at a premium solenoidal coils are preferable to saddle-shaped coils as long as any consequent reduction in B_0 homogeneity can be tolerated and as long as the physiology does not suffer. The reduction in field homogeneity can certainly be tolerated for most ^{31}P and [13]C n.m.r. studies of living systems. This general conclusion regarding the relative sensitivity of the two types of coil requires slight modification when studying large physiological samples because the electrical conductivity of these samples generates an additional source of noise which must be taken into account. This effect is discussed in § 8.5.

8.4.2. Surface coils

It is routine practice in n.m.r. experiments to place the sample *within* the volume contained by the radiofrequency coil. This, of course, makes sense because it is within this volume that the B_1 field is the greatest and it is therefore from this volume that most of the signal is detected. However, for studies of living systems it is not always convenient or even possible to surround the sample by a radiofrequency coil. For many years flat 'surface' coils have been used to monitor blood flow (see for example Morse and Singer 1970). However, the quality and versatility of these coils was not widely appreciated, and it was not until recently that they were used with great success to monitor the metabolism of live animals. The simplest form of surface coil is simply a loop of wire, as shown in Figs. 8.1 and 8.2(a). If such a coil is placed adjacent to a large sample, most of the signal induced in the coil will in general originate from an approximately disc-shaped region of the sample immediately in front of the coil.

Surface coils have three particularly useful properties. Firstly, the signal-to-noise ratio that can be obtained is similar to that obtained using more conventional coils. Secondly, surface coils can be used to obtain information about the spatial distribution of compounds close to the surface of a sample. Thirdly, they can be made as large or as small as desired regardless of the total size of the sample. Furthermore, surface coils need not necessarily be circular in cross-section; their geometry can be designed to suit the shape of the sample. Experiments performed

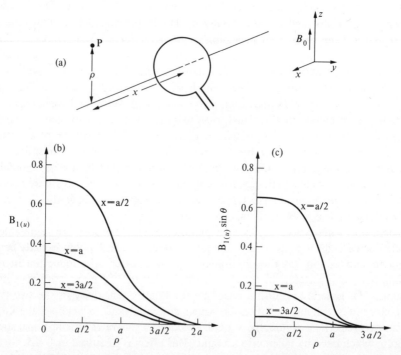

Fig. 8.2. (a) Diagram illustrating a circular surface coil oriented within the yz-plane. A point in space P is shown, described by the coordinates (ρ,x). For this particular point, ρ is parallel to B_0. (b) The spatial dependence of $B_{1(u)}$, and (c) the spatial dependence of $B_{1(u)} \sin \theta$, plotted as a function of ρ, the radial coordinate, and x, the axial coordinate. ρ and x are expressed in units of the coil radius a, and $B_{1(u)}$ is normalized to 1.0 at the point $(\rho,x) = (0,0)$; i.e. at the centre of the coil. At this point, the pulse angle θ is $90°$. If ρ were not parallel to B_0, the spatial variations would be slightly different (see Ackerman *et al.* 1980*b*).

with these coils therefore possess a a great deal of flexibility with regard to sample size and shape. For these reasons it is undoubtedly true that surface coils will become increasingly used for studies of living systems.

Surface coils work in exactly the same manner as conventional solenoidal or saddle-shaped coils, the main distinction being that the B_1 field that surface coils generate within a sample is not as uniform as that produced by the more conventional coils. Because the B_1 field is not homogeneous throughout the sample volume, the expression for the signal-to-noise ratio given by eqn (8.2) must be modified. In assessing the signal generated by different parts of the sample, it is important to note that regions where the B_1 field is small produce weak signals for two reasons.

(i) If the radiofrequency pulse length corresponds to a $90°$ pulse in the region where the $B_{1(u)}$ field is strongest, it will correspond to a much smaller pulse angle in regions of weak $B_{1(u)}$ field. This is because the pulse angle θ is equal to $\gamma B_{1(u)} I t_p$ where t_p is the length of the pulse and I is the current passing through the coil. (This expression is equivalent to eqn (7.1)). Therefore the magnetization of the sample will be smaller in regions of weak field.

(ii) Even if the magnetization following a radiofrequency pulse were uniform throughout the sample, those regions of the sample where the $B_{1(u)}$ field is weak would generate less signal than those regions where it is large. In fact, there would be a linear dependence of signal on $B_{1(u)}$ (see eqn (8.2)).

The signal from a given volume element is therefore proportional to $B_{1(u)} \sin \theta$, and eqn (8.2) must be modified to the more precise form

$$\epsilon = \omega_0 M_0 \int B_{1(u)} \sin \theta \, d V_s \qquad (8.5)$$

It should be noted that if the pulse angle is chosen to be $90°$ in the region of strongest radiofrequency field, the spatial distribution of $B_{1(u)} \sin \theta$ is far more strongly defined than that of $B_{1(u)}$. This can be seen from Fig. 8.2, which shows the spatial distribution of $B_{1(u)}$ and $B_{1(u)} \sin \theta$ that is produced under these conditions by a circular surface coil. It can be seen from Fig. 8.2(c) that under normal experimental conditions the signal will be predominantly from a disc-shaped region of the sample which is immediately in front of the coil and which has a radius and thickness approximately equal to the radius of the coil. Moreover, the noise generated by a conducting sample (see § 8.5) also originates in this region, because this is the only region of the sample where the B_1 field is strong. Therefore, an important characteristic of surface coils is that only that part of the sample that contributes signal also contributes noise. This contrasts sharply with focusing experiments in which a large coil surrounds the whole sample but detects signal from only a small localized region of the sample. Under those conditions, the whole sample contributes noise, but a much smaller volume contributes signal. Surface coils therefore provide a very efficient means of detecting signal from localized regions, as long as these regions lie close to the surface of the sample.

Of course, the pulse angle θ can be chosen to suit the experimental situation, and in principle it should be possible to saturate selectively the region of the sample closest to the coil and to enhance the relative contribution from other regions further from the coil. However, the signals obtained in this way would originate from a very diffuse region of the sample and moreover would be relatively weak. This approach alone is therefore not suitable for obtaining signals from deep within a sample. However, when used in conjunction with other methods of localization, there could be some interesting possibilities. For example, in experiments designed to localize on the kidney of live rats (Balaban *et al.* 1981*a*) there is a problem in that a layer of muscle between the kidney and the surface coil generates strong signals. However, these signals can be almost completely removed by ensuring that the layer of muscle experiences pulses of angle about 180°. The kidney experiences pulses of smaller angle because of the B_1 field gradient associated with the surface coil and therefore produces reasonably strong signals. The signals from the outlying regions can be removed by means of the B_0 field profiling of topical magnetic resonance (see § 4.1.2), and so the resulting spectrum is almost exclusively derived from the kidney.

8.5. The effects of conducting samples

Living systems have fairly high concentrations of ions, and as a result they conduct electricity. The electrical conductivity of tissues and organs is approximately equal to that of a solution of 100 mM NaCl and for large samples this can have profound consequences. In particular, the sample itself can generate a significant amount of noise, and this has to be taken into account when considering signal-to-noise ratios. In addition, there will be a limit to the depth of penetration of the radiofrequency field into the sample. Let us firstly consider the signal-to-noise ratio, assuming for the present that the B_1 field is uniform throughout the sample volume.

8.5.1. Magnetic losses

Following a radiofrequency pulse, the magnetization developed in the xy-plane induces an e.m.f. in the receiving coil. A back e.m.f. is also induced in the sample itself, and if the sample is conducting, currents flow within it and these so-called eddy currents dissipate power. This power loss is analogous to power losses within the resistance of the coil itself and therefore represents an additional source of resistance R_m. The subscript m refers to the fact that the resistance is associated with the magnetic field of the radiofrequency coil, and the expression 'magnetic losses' can be used to describe the power dissipation.

The total effective resistance R_t can now be written

$$R_t = R_c + R_m \tag{8.6}$$

where R_c refers to the inherent resistance of the coil. The signal-to-noise ratio ψ is therefore proportional to $B_{1(u)} V_s/(R_c + R_m)^{1/2}$. If we also include frequency as a variable, then we find from eqns. (8.2) and (8.3) that

$$\psi \propto \omega_0^2 B_{1(u)} V_s/(R_c + R_m)^{1/2} \tag{8.7}$$

It can be shown that, for a cylindrical sample enclosed by a solenoidal coil,

$$R_m = V_s^2 B_{1(u)}^2 \omega_0^2 \sigma / 16 \pi g \tag{8.8}$$

(see Hoult and Lauterbur 1979), where σ is the conductivity of the sample and $2g$ is the length of both sample and coil. Therefore by combining eqns (8.7) and (8.8) we find

$$\psi \propto \omega_0^2 B_{1(u)} V_s / \{R_c + [B_{1(u)}^2 V_s^2 \omega_0^2 \sigma / 16 \pi g]\}^{1/2} . \tag{8.9}$$

From this equation it is clear that if we wish to increase the signal-to-noise ratio at a given frequency, we must increase the product $B_{1(u)} V_s$ that occurs in the numerator. However, as $B_{1(u)} V_s$ increases, the value of R_m increases rapidly and eventually it will become larger than R_c. When this occurs, we can write

$$\psi \propto \omega_0^2 B_{1(u)} V_s / R_m^{1/2} . \tag{8.10}$$

From eqn (8.8)

$$R_m^{1/2} = B_{1(u)} V_s \omega_0 \sigma^{1/2} / 4 \pi^{1/2} g^{1/2} \tag{8.11}$$

and therefore

$$\psi \propto \omega_0 g^{1/2} \sigma^{-1/2} . \tag{8.12}$$

i.e. ψ is independent of $B_{1(u)}$ and R_c, and depends upon the variables ω_0, g, and σ. Other parameters in the full expression for ψ are beyond experimental control. This important observation leads to the following conclusions.

(i) For a given sample shape and coil configuration, if the condition is reached that $R_m > R_c$ any improvement in coil design will have very little effect on the signal-to-noise ratio. Therefore, in striving to optimize signal-to-noise ratios we should attempt to approach this limiting condition, and on reaching this condition there is little point in continuing further. A corollary of this is that a saddle-shaped coil should be almost as effective as a solenoidal coil if $R_m > R_c$. However, it should be noted that the expression for R_m given in eqn (8.8) was derived for a solenoidal coil; the corresponding value for R_m when using a saddle-shaped coil would differ. (This difference arises because $B_{1(u)}$ is directed perpendicular to rather than along the axis of the sample when a saddle-shaped coil is used.) The general conclusion that can be reached is that when $R_m > R_c$ a solenoidal coil should still have slightly better sensitivity than a saddle-shaped coil; however, the difference might be so small that the choice of coil design is governed by factors other than sensitivity.

(ii) Once the condition $R_m > R_c$ has been reached, the signal-to-noise dependence on sample volume becomes very weak; in the case of a cylindrical sample enclosed by a solenoidal coil the signal-to-noise ratio is independent of the sample radius and depends only on the square root of length. There is therefore little to be gained by increasing the sample volume.

(iii) The ultimate signal-to-noise ratio that can be achieved is proportional to the square root of sample conductivity and is linearly related to resonance frequency. This dependence on frequency is weaker than when sample losses are unimportant and so the gain in sensitivity obtained on increasing magnetic field strength may be rather disappointing.

In view of these conclusions, it is essential to assess the relative values of R_c and R_m that we might expect to obtain. A well-designed radiofrequency coil typically has a resistance of about $0.1 - 0.5 \, \Omega$ and it is instructive to calculate an approximate value of R_m for a typical experiment. Equation (8.8) can alternatively be written

$$R = \pi \omega_0^2 n^2 f^4 \mu_0^2 g \sigma / 16 (a^2 + g^2) \tag{8.13}$$

where n is the number of turns in the coil, a is its radius, $2g$ is the length of both coil and sample, and f is the radius of the sample. If we put $a = g$ and $f = 0.75a$, then we find that at 70 MHz, R_m is approximately $0.15 \, \Omega$ for a five-turn coil of radius 1 cm. For a larger sample, we might use a three-turn coil of radius 2 cm. It is then found that R_m is about $0.45 \, \Omega$.

It seems that, for a variety of studies of living tissues and organs, R_m and R_c may both be in the range $0.1-0.5 \, \Omega$. Under these conditions a solenoidal coil should still produce better sensitivity than a saddle-shaped coil, but we can anticipate that the difference will be by a factor of around 1.5 and will probably be less for samples of volume 10 ml or greater. Clearly, the numbers depend upon the operating frequency of the spectrometer, the nucleus under investigation, the sample shape etc., but these figures should provide rough guidelines for those interested in coil design.

The most frustrating aspect of all this is that the magnetic losses seem to impose an upper limit on the sensitivity that can *ever* be attained. There appears to be little hope of ever enhancing sensitivity by a factor of, say, 10 for studies of *large* living systems, and we must resign ourselves to the view that for these studies the signal-to-noise ratios that can presently be obtained on well-designed spectrometers are not very far away from the maximum that can be achieved. However, this does not mean that further developments in coil design deserve little further attention. A very wide range of n.m.r. studies are performed on samples of very low conductivity, and the sensitivity of such experiments would certainly benefit from improved coil design. In addition, improved coil design would provide better sensitivity for *small* conducting samples (the dependence of R_m on the volume should be noted), and therefore the possibility does arise that future experiments may be feasible on much smaller samples than can presently be studied. One possible future development is to reduce the temperature of the coil (insulating it, of course, from the sample) and thereby reduce its resistance. However, there are considerable engineering and electronic problems associated with reducing the coil temperature to very low values (e.g. liquid-helium temperature).

It should be noted that the magnetic losses from a sample would be considerably reduced if it were possible to place within it some insulating material that would interrupt the current paths. This should be feasible for cell preparations, but would clearly be impractical for studies of a single tissue or organ. It should also be noted that the sensitivity of an n.m.r. experiment would be improved if it were possible to enhance the magnetization of the sample by some means. Ingenious pulse methods have been devised for enhancing the signal-to-noise ratio for insensitive nuclei (for example the INEPT sequence of Morris and Freeman (1979)) and it is conceivable that such methods may provide an alternative means of enhancing signal-to-noise ratios in studies of living systems.

8.5.2. Dielectric losses

So far, we have considered only the magnetic properties of radiofrequency coils. However, there are also unavoidable electric fields associated with the coils, and if electric lines of force pass through a conducting sample, these provide an additional loss mechanism. Just as with magnetic losses, this can be expressed in terms of an effective resistance R_e where the subscript e indicates that the losses are electrical in origin.

There is no point in attempting to reduce the magnetic losses by reducing the magnetic fields with which they are associated, because by so doing the signal would also be reduced. However, there is everything to be gained in preventing electric fields from entering the sample, as long as the preventive measures that are taken do not in themselves introduce additional losses. The procedure that is adopted, when necessary, is to place a Faraday shield (see, for example Gadian and Robinson 1979) between sample and coil, and if designed correctly this can prevent electric fields from entering the sample. However, it may be difficult to ensure that the shield does not introduce additional losses that reduce its value.

It has been found for the probes built in the Oxford laboratory that a Faraday shield is very rarely of significant benefit, although a shield was used for a seven-turn solenoidal coil in which the turns were closely spaced. In practice, coils can usually be designed in such a way that dielectric losses are small in comparison with magnetic losses and/or inherent losses within the coil itself. The interested reader is referred to the articles by Hoult and Lauterbur (1979) and Gadian and Robinson (1979) for further discussion of the topic.

8.5.3. Radiofrequency field penetration

A variety of important effects emerge when we consider the interaction of electromagnetic fields with conducting materials. One of these effects has already been discussed, namely the dissipation of power within conducting samples which introduces additional noise into the n.m.r. receiver. It can also be shown that, at high frequencies, the electromagnetic field and the current within a conductor are both confined to the outer surface of the conductor. This phenomenon is known as the skin effect, and in n.m.r. has two important consequences. Firstly, the current flowing in the radiofrequency coil is confined to the outer surface of the wire or foil, which is an important factor to be considered when assessing coil resistance (see Appendix 8.1). Secondly, the radiofrequency field B_1 has a limited depth of penetration into conducting samples, the depth decreasing according to the square root of the resonance frequency. This generates an upper limit to the sample size that can generate useful n.m.r. signals, and detailed calculations of this effect have been performed by Bottomley and Andrew (1978). The skin effect in conducting samples leads to distortion of both the amplitude and the phase of the magnetic field, and Bottomley and Andrew concluded that for human body studies this will generate problems at frequencies above about 30 MHz. At 100 MHz there should be few problems for samples of radius 10 cm or less. Fortunately, therefore, there will be few experimental situations in which finite magnetic field penetration provides a limitation to useful sample size.

8.6. The tuned circuit

Power is transferred to and from the probe via a coaxial cable, and efficient power transfer only takes place if the electrical characteristics of the coil circuitry match those of the cable. Cables usually have a so-called 'characteristic impedance' of 50 Ω, and it is therefore essential for the probe circuit to appear to behave as a 50 Ω resistance. The characteristics of the coil itself are very different from this, and therefore we cannot simply connect the coil across the end of the cable. Instead, its characteristics must in some way be transformed to the required 50 Ω, and this is most conveniently achieved using a circuit of the type shown in Fig. 8.3 (see Appendix 8.1 for an explanation of the components of the circuit and the symbols that are used).

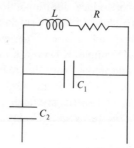

Fig. 8.3. A tuned circuit commonly used in n.m.r. probes.

The tuning capacitance C_1 and the coil, which has an inductance L and resistance R, constitute a tuned circuit which has a resonance frequency ω_0 given by

$$\omega_0^2 = 1/LC_1 . \tag{8.14}$$

At this frequency a particularly large magnetic field can be generated for a given input power. Therefore by ensuring that the coil resonates at the resonant frequency of the nuclei we can obtain the large B_1 field that is required in pulsed Fourier transform n.m.r. However, the tuning capacitance in itself does not ensure efficient power transfer because the circuit will still not appear to behave as a 50 Ω resistance. What is required is an additional 'matching' capacitance C_2 as shown in the figure. It is not intuitively obvious that a circuit such as this can be equivalent to a 50 Ω resistance, but nevertheless it works; the logic behind the circuit is discussed by Hoult (1978). The addition of the capacitors does not affect the signal-to-noise ratio arguments presented above, provided that they are of high quality and introduce no further noise into the circuit. The reason for this is that the two capacitors basically function as a transformer, and they transform signal and noise to precisely the same extent.

Prior to starting an experiment, it is essential to ensure that the probe is tuned and matched, i.e. that the values of C_1 and C_2 are such that the probe behaves like a resistance of 50 Ω at the resonance frequency. Probe tuning can be performed with the aid of a radiofrequency bridge circuit such as that described by Hoult

(1978), and is achieved by adjusting C_1 and C_2 until a null output is observed from the bridge. Although the two capacitors interact with each other, a simple iterative procedure usually enables the null to be readily obtained. It is useful to note that for a high-quality tuned circuit C_2 should be smaller than C_1 by a factor of about 5 or more, and that this factor decreases as the quality of the circuit declines. As we have seen in previous sections, the effective resistance of the radiofrequency coil, and hence the required values of C_1 and C_2, depend on a number of factors, including the conductivity of the sample, and therefore it is not inconceivable that the tuning will change on replacing one sample by another. When using small samples, e.g. in conventional high-resolution ^{1}H n.m.r. experiments, it can generally be assumed that the tuning will not change from one sample to another, but for the type of experiment described in this book it is imperative to check, and if necessary to adjust, the tuning prior to each experiment. Incorrect tuning will not only degrade the signal-to-noise ratios, but will also reduce the strength of the B_1 field causing the 90° pulse length to increase.

Some practical details of tuned circuits used in n.m.r. probes are given in § 8.8.

8.7. Some practical details of coil design

8.7.1. Sensitivity

Suppose that it has been decided to use a solenoidal coil of specified overall dimensions for a given experiment. The coil is to be made from copper wire, because copper has high conductivity and can readily be obtained at high purity (any paramagnetic impurities might reduce the homogeneity of the field B_0). We then have to decide how many turns the solenoid should contain and how thick the wire should be.

If we remember that the ratio $B_{1(u)} V_s / R^{1/2}$ should be maximized, it might seem that a large number of turns should be used, as we might expect $B_{1(u)}$ to increase according to the number of turns but $R^{1/2}$ to increase according to the square root of this number. However, as the number of turns increases, the radius of the wire must decrease in order for it to fit into the same space, and therefore $B_{1(u)} V_s / R^{1/2}$ might not in fact increase.

Additional factors also argue against using a very large number of turns. Firstly, when the turns are closely wound, the coil resistance is increased as a result of an effect known as the 'proximity effect' (see Hoult 1978). Perhaps more importantly, a stage is reached when the coil approaches 'self-resonance'. This describes the situation when the coil resonates in the absence of a tuning capacitance; its own 'self-capacitance' acts as the tuning capacitance. Under these conditions the performance of the coil deteriorates considerably. Moreover, with conducting samples dielectric losses become particularly severe close to self-resonance. Therefore it is important to ensure that the coil is not close to self-resonance, i.e. the tuning capacitance must be much greater than the self-capacitance of the coil, which may typically be about 1 pF. If possible, therefore, the tuning capacitance should have a value of about 8 pF or greater, and the number of turns in the coil should be selected accordingly. Hoult (1978) suggested that to minimize the resistance, the separation of the turns should be about three times the wire radius, and by using these criteria it should be possible to select the number of turns and wire radius that produce a performance close to optimal.

As the resonance frequency and coil size increase, it becomes increasingly difficult to design coils that do not self-resonate. However, two coil designs that do operate well at high frequencies are shown in Fig. 8.4; essentially, they are low-inductance solenoidal and saddle-shaped coils.

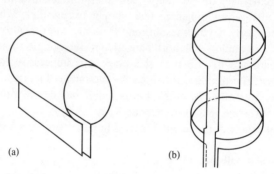

(a)   (b)

Fig. 8.4. (a) A low inductance solenoidal coil, and (b) a low inductance saddle-shaped coil, due to Dadok.

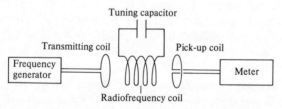

Fig. 8.5. Schematic diagram illustrating the apparatus that can be used for predicting the sensitivity of radiofrequency coils.

The sensitivity of radiofrequency coils can be tested very simply using the apparatus illustrated schematically in Fig. 8.5. A transmitting coil, connected to a source of variable frequency, is placed fairly close, but not too close, to the radiofrequency coil across which there is a tuning capacitor. A pick-up coil is placed on the other side of the tuned circuit and is connected to a meter or oscilloscope. When the tuned circuit is far away from resonance no signal should appear on the oscilloscope. However, as the variable frequency approaches resonance currents flow in the tuned circuit, the effects of which are detected by the pick-up coil and a signal should be observed. The intensity of this signal plotted as a function of frequency should be Lorentzian in shape, just like an n.m.r. signal, and it can be shown that the 'quality factor' Q of the coil (see Appendix 8.1) is given by the ratio of the signal frequency to the bandwidth at $1/\sqrt{2}$ of the peak amplitude. The presence of the transmitting and pick-up coils close to the tuned circuit can reduce the measured Q. Control experiments should be performed to ensure that this reduction in Q is negligible by moving the coils further away from the tuned circuit and checking that the measured Q stays the same.

To determine the characteristics of a radiofrequency coil a capacitance of known value should firstly be placed across the coil and the resonance frequency should be measured using the above procedure. Since $\omega_0^2 = 1/LC$ and C is known, the inductance L of the coil can be evaluated. Then, by using a variable high-Q capacitance the resonance frequency of the circuit should be adjusted to be equal to the nuclear resonance frequency, and the value of Q should be measured at this frequency. Since $Q = \omega_0 L/R$ and both Q and L are known, the resistance R of the coil can be estimated. The $B_{1(u)}$ field of solenoidal and saddle-shaped coils can be estimated from their geometry (see Hoult and Richards 1976). Hence the signal-to-noise ratio, which is proportional to $B_{1(u)} V_s/R^{1/2}$, can be evaluated. By means of these relatively simple experiments, the sensitivity of radiofrequency coils can be assessed without any requirement for a spectrometer; all that is needed is a frequency source and a meter or oscilloscope. Alternatively, a device known as a Q-meter can be used.

8.7.2. B_0 and B_1 homogeneity

It is difficult to be specific about B_0 and B_1 homogeneity, but a number of general considerations can be pointed out.

It is very likely that thick copper wire will adversely affect B_0 homogeneity, particularly if the wire is close to the sample. It may cause 'wings' in the spectral lines, the broad underlying component arising from that region of the sample that is closest to the wire. Although this may be tolerable for studies of living systems, it can produce difficulties in certain experiments. For example, in ^{1}H n.m.r. it is usually essential to saturate the solvent peak in order to observe clearly the much smaller resonances from metabolites etc. If there is a broad component to the solvent peak, it will be impossible to eliminate this without also eliminating some of the signals of interest. For such experiments it would therefore be necessary to employ thinner wire or preferably foil.

Perhaps the main point, however, is that the great care that is taken to ensure that optimal homogeneity is preserved for very-high-resolution experiments is somewhat misplaced for a wide variety of studies of living systems. When the ultimate in B_0 homogeneity is not essential, we can take many more liberties than might have been anticipated. Therefore, the approach should be to design a probe with optimal sensitivity and simply try it out to see whether the resulting B_0 homogeneity is satisfactory.

It is difficult to provide a theoretical assessment of the B_1 homogeneity of different coil designs because of uncertainties about the current paths within a wire or foil due to imperfections in coil winding etc. It is certainly true that for those n.m.r. studies in which really good B_1 homogeneity is required it is common practice to use a transmitting coil that is much larger than the sample. This could also be used as the receiving coil if sensitivity is not a problem, but often a second receiver coil is used which is smaller and therefore more closely coupled to the sample. However, for studies of living systems such a procedure is cumbersome, is bound to be less sensitive than using a closely coupled single coil, and is almost certainly unnecessary.

Some precautions can generally be taken to reduce B_1 homogeneity problems. For example, the B_1 homogeneity of a saddle-shaped coil is optimal when the arc angle of each turn is $120°$ (see Hoult 1978). More generally, the sample should not extend too close to the coil where the B_1 field is certainly inhomogeneous; as a very rough guide the sample radius should be no more than about 0.75 times the coil radius. This is also advantageous from the point of view of B_0 homogeneity, but of course the figure of 0.75 must depend greatly on the thickness of the wire that is used. In addition, the B_1 field of a solenoidal coil extends axially considerably beyond the confines of the coil, gradually declining with distance. Therefore, if B_1 homogeneity is important, the sample length should not be greater than the coil length.

A most important point is that the B_1 field is very large in a small region immediately adjacent to the radiofrequency coil. Therefore what constitutes a $90°$ pulse for the majority of the sample would correspond to a much larger pulse angle for this small region. Of course, this region would not generate large signals under these circumstances, because its volume is so small and because the nuclei within it are saturated; however, a conducting sample in this region would generate a lot of noise because the eddy currents within this region would be very large. As a result, it is essential when using surface coils to ensure that there is a distance of a millimetre or thereabouts between the coil and any conducting part of the sample. It is advisable for the coil to be well insulated so that there is no chance of any moisture getting close to the copper.

A related point is that adjustment of the B_0 homogeneity is often performed using the ^1H signal from water within the sample. The concentration of water is so great that sufficient signal can often be detected with a coil that is not tuned to the ^1H frequency (Ackerman *et al* 1981). The simplest procedure is to employ the radiofrequency coil that is used for detecting the nucleus of interest, whether this be ^{31}P or any other nucleus. However, because the coil is not tuned to the ^1H frequency, the B_1 field at the ^1H frequency will be very small and so the radiofrequency pulses will be far less than $90°$ for the bulk of the sample. Suppose that throughout most of the sample volume, the pulse angle is about $1°$. In the small region immediately adjacent to the coil the angle will be much greater, but it is now unlikely that the nuclei in this region will be saturated. As a result, they may generate the predominant part of the total signal. Shimming of the B_0 field may therefore be inadvertently performed on a very small volume close to the coil rather than on the bulk volume of the sample, the homogeneity of which may remain very poor. This provides another important reason why the sample should not be positioned immediately adjacent to a radiofrequency coil; there should be a small space between coil and sample.

8.8. Some examples of probes

In this section we describe some probes based on those currently used in the Oxford laboratory. These probes are designed for use on a spectrometer equipped with a wide-bore (about 9 cm) superconducting magnet operating at a field of 4.3 T. Although some of the precise details may not be of direct relevance to probes for use in other laboratories, nevertheless the underlying principles and problems will be common to all types of spectrometer.

The probe body is a cylindrical tube made of a conducting material such as aluminium. It functions as a shield for the radiofrequency circuitry, preventing the leakage of stray radiation into the probe. The bodies built in the Oxford laboratory are made of an aluminium alloy, and have two detachable halves, permitting easy access to the radiofrequency coil and sample when setting up an experiment. This can be seen from Fig. 1.11. Since the basic design of the probe body is common to a variety of different probes, it is economical to build inserts specifically designed to suit a particular type of experiment. The complete probe therefore consists of a fixed body and a variable insert, and on changing from one type of experiment to another it is simply necessary to replace one insert by another.

8.8.1. The basic electronics

In the simplest situation, e.g. a ^{31}P n.m.r. experiment that requires no irradiation at a second frequency, the radio-frequency tuned circuit consists of the coil and two capacitors as was shown in Fig. 8.3. The requirements for coil design have been discussed earlier in this chapter; here, therefore, it is only necessary to discuss the other aspects of the tuned circuit.

The capacitors should have high Q values, or quality factors, to ensure that they do not introduce any additional losses into the tuned circuit. (It should also be noted that the circuit boards on which the capacitors may be mounted are often themselves lossy, so they should be used with appropriate caution.) The capacitors should be able to withstand the high voltages (typically about 1 kV) used in Fourier transform n.m.r., and they should be made of non-magnetic materials. The tuning capacitor should be positioned as closely as possible to the radiofrequency coil in order to minimize resistive losses in the leads to the coil. If it is inconvenient or impractical to place the variable capacitor very close to the coil, the losses can be reduced by having a fixed tuning capacitor close to the coil in parallel with a further variable tuning capacitor. Of course, if a capacitor is placed close to the coil, care has to be taken to ensure that it does not appreciably disturb the homogeneity of the static field B_0. In addition, it should be ensured that the electric fields associated with the capacitor do not pass through the sample, as these could generate dielectric losses (see § 8.5.2).

The earth connection to the circuit can be made by means of a strip of copper foil that is attached to the probe body, and the tuned circuit is connected to the outside world by a length of coaxial cable of characteristic impedance 50 Ω which is earthed at the base of the probe body. A cable from the probe base leads to the pre-amplifier, which constitutes the first stage of the spectrometer receiver.

For many experiments, e.g. ^{13}C studies for which proton decoupling and/or Overhauser enhancements are essential, an additional coil tuned to the ^{1}H frequency is required. In principle, it is possible to double-tune a single coil so that it is responsive to two frequencies, but for ^{13}C studies this is undesirable as double-

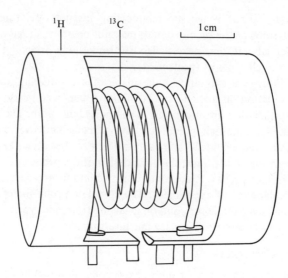

¹H ¹³C   1 cm

Fig. 8.6. The radiofrequency coil design that is used for ¹³C n.m.r. studies of living systems.

tuning invariably leads to a reduction in the signal-to-noise ratio at the lower frequency. The radiofrequency coil design for a ¹³C probe constructed in the Oxford laboratory is illustrated in Fig. 8.6. Care is taken to ensure that the radiofrequency fields produced by the two coils are perpendicular to each other, for otherwise the two coils would interact excessively with adverse effects on their tuning. It should be noted that the ¹H coil is of the type shown in Fig. 8.4(b) and is effectively a saddle-shaped coil with the two halves connected in parallel. The reason for this is that this type of coil has a very low inductance for its size and can therefore be easily tuned at about 180 MHz. A conventional saddle-shaped coil of this size would be close to self-resonance and would perform poorly.

8.8.2. A skeletal muscle probe

A sample chamber that has been used for studies of skeletal muscle is illustrated in Fig. 8.7. The saddle-shaped radiofrequency coil is made from wire of radius 1 mm and is self-supporting. The main reason for using a saddle-shaped rather than a solenoidal coil is that muscle tension can be measured much more easily when the muscles are held vertically than when they are held horizontally. The sample tube is simply a glass vial of inner radius about 6 mm, costing about 1 p or 2 cents! Homogeneity of 1 part in 10^9 cannot be expected with such a configuration, but linewidths of about 2 Hz at 73 MHz can readily be achieved with an unspun sample, and this is more than adequate for ³¹P studies of living systems.

The vial is fitted with a plastic stopper at the lower end, through which is placed a hook made of tinned copper which acts as both a stimulating electrode and a support for the muscles. Four or more frog gastrocnemius muscles, weighing about 2 g *in toto*, are mounted vertically within the vial, and threads attached to their

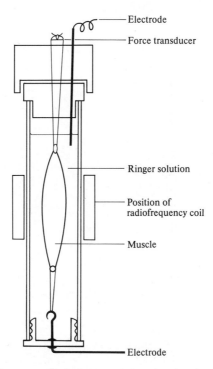

Fig. 8.7. The main features of an n.m.r. probe that has been used for studies of skeletal muscle by Dawson, Gadian, and Wilkie.

tendons are tied at the lower end to the hook and at the upper end to a force transducer described by Dawson *et al.* (1977*a*). Radiofrequency stoppers are placed in the stimulating circuit immediately above and below the chamber to minimize the losses generated by the stimulation leads. Each stopper consists of a parallel tuned circuit which is tuned to the resonance frequency of the nuclei under investigation. This generates a high resistance at the resonance frequency but a relatively low resistance at the low (about 50 Hz) frequency that is used for the electrical stimulation.

Temperature within the probe is monitored with a copper-Constantan thermocouple. Because it is difficult to avoid fairly large temperature gradients in a probe of this type, the thermocouple is positioned immediately adjacent to the vial (but not within the field of the radiofrequency coil for if it is placed there it would interfere severely with the radiofrequency circuitry). Because of this positioning, it is particularly important to place radiofrequency chokes or stoppers in the thermocouple leads to prevent excessive leakage of stray radiation into the probe. These must also be placed in the electrical leads to the tension transducer, for the same reason. The adverse effects of leakage are discussed in § 7.3.1.

The muscles are bathed in Ringer's solution, which can if necessary be pumped in and out of the vial. The tubing that transports the solution to and from the vial must be held well away from any parts of the radiofrequency circuitry because the

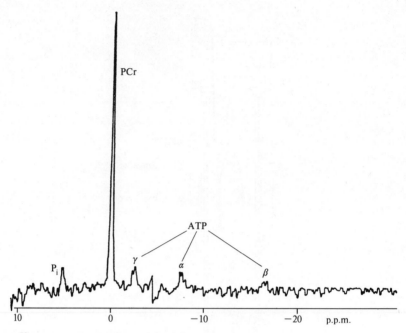

Fig. 8.8. ^{31}P n.m.r. spectrum obtained at 73.8 MHz from about 1.5 g of muscle, using a single 90° pulse. (Adapted from Dawson *et al.* 1980*b*.)

presence of conducting material close to the tuning capacitor, for example, would generate losses and therefore reduce the quality of the tuned circuit.

For experiments in which an optimal signal-to-noise ratio is the prime considera-tion, the saddle-shaped coil is replaced by a five-turn solenoidal coil and the vial is placed horizontally within the solenoid. This leads to a 50 per cent improvement in the signal-to-noise ratio, so that the time required for each spectral accumulation is halved. Loss of B_0 homogeneity is found to be insignificant, and Fig. 8.8 shows a spectrum obtained from about 1.5 g of muscle using a single 90° pulse.

An earlier version of the muscle chamber that was used with a narrow-bore magnet operating at 7.5 T is discussed in detail by Dawson *et al.* (1977*a*).

8.8.3. A probe for studying perfused hearts

A probe for perfused heart experiments was shown in Fig. 1.11. In addition to the types of problems encountered with the skeletal muscle studies, a number of addit-ional features can be pointed out.

In order to ensure that the field B_0 is as homogeneous as possible, the number of air—liquid interfaces should be minimized. Therefore the heart is immersed in buffer and care is taken to ensure that there are no bubbles in the sample tube (but see § 8.8.6 for further discussion of bubbles). Thus the perfusion medium is oxy-genated outside the spectrometer and passes through a bubble trap prior to entering the probe. It is advisable to add a small amount of ethylenediaminetetraacetic acid (EDTA) to the perfusion medium to chelate any paramagnetic ions, and in this respect it should be noted that cleaning glassware in chromic acid can have disastrous broadening effects on spectral lines!

The volume of buffer enclosed by the coil should be kept small to minimize the noise generated by the sample (see § 8.5). However, the sample tube must be large enough to ensure that the heart can beat freely without hitting the walls of the tube, for even light contact can cause ischaemic patches to appear. The cannulae for perfusion are made of glass, rather than stainless steel, to avoid radiofrequency losses that they may produce. Further details of the apparatus used for heart perfusion are given by Garlick *et al.* (1979).

Pacing of hearts poses particularly acute problems, because the pacing electrodes must be positioned directly against the heart within the volume enclosed by the radiofrequency coil. It is difficult to avoid the consequent reduction in the signal-to-noise ratio; radiofrequency stoppers should be placed in the electrode leads as closely as possible to the sample tube but these will not totally eliminate the losses. One way of reducing the losses is to keep the amount of metal in the electrodes to a minimum and to make electrical contact from the leads to the metal with salt bridges. Salt bridges have been used by Hollis *et al.* (1978) in their studies of perfused hearts. They describe in their paper how such studies can be performed with the use of relatively minor modifications to the conventional n.m.r. probes.

8.8.4. A probe for studies of perfused liver

For studies of perfused liver, the glass sample tube has been replaced by a chamber made of Perspex (see Fig. 8.9). The Perspex is hollowed out to provide an approximately liver-shaped cavity, and the radio-frequency coil that fits around the chamber is shaped like an earphone. Excellent signal-to-noise and spectral resolution can be obtained with this design (see Iles, *et al.* 1980).

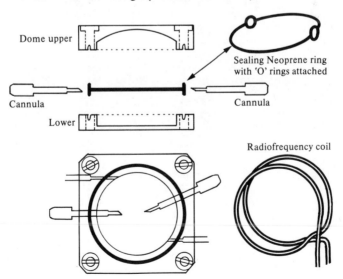

Fig. 8.9. A sample chamber that has been used for studies of perfused liver. (From Iles, *et al.* 1980.)

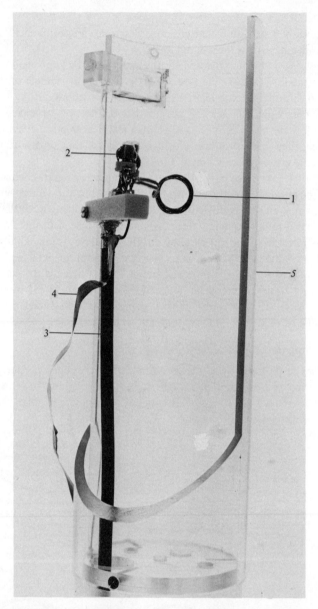

Fig. 8.10. The main features of a surface coil probe built in Oxford by K. Thulborn for studies of anaesthetized animals. 1, the surface coil; 2, tuning and matching capacitors; 3, coaxial cable leading to and from the tuned circuit; 4, earth strap; 5, animal cradle.

8.8.5. A surface coil probe

In many ways this is much simpler than the probes described above. The more conventional type of coil is replaced by a surface coil of the required dimensions, and in the probe illustrated in Fig. 8.10 the coil can be placed adjacent to the brain or muscle of an anaesthetized rat. Again, unnecessary losses must not be introduced into the system. Therefore the coil should be carefully insulated, the capacitors should be placed close to the coil, and the coil should not be placed too close to the sample (see § 8.7.2). The leads to the coil should be positioned in such a way that they themselves do not detect signal, for this could greatly confuse interpretation of the spectra and adjustment of the field homogeneity.

8.8.6. Some general points

Bubbles within the sample can severely impair B_0 homogeneity, and this can create problems if oxygen is to be introduced into a sample. Navon, Ogawa, Shulman, and Yamane (1977*c*) have described a method for overcoming this problem in their studies of cell suspensions. They used the deuterium lock channel to detect the presence of bubbles, and were therefore able to synchronize data collection with the periods between successive bubblings. Moreover, Ugurbil, Shulman, and Brown (1979) commented that a continuous stream of oxygen bubbles can be used for oxygenation of *E. coli* suspensions without causing significant magnetic field inhomogeneities provided that the bubbles are small and that the flow is regular. However, these procedures are far from ideal, and a sample chamber has now been developed for maintaining adequate oxygenation of cellular suspensions without introducing bubbles (Balaban *et al.* 1981*b*).

Oldfield and Meadows (1978) have described a sideways spinning probe in which they used a solenoidal coil to optimize the signal-to-noise ratio and also spin the sample to improve the spectral resolution. This type of probe could be most useful for ^{13}C and ^{31}P studies of solutions. However, for ^{1}H studies, it remains probable that the best approach is to use a fairly conventional type of probe. The reason for this is that lack of sensitivity is not so much of a problem in ^{1}H n.m.r. Moreover, really good homogeneity is generally required for ^{1}H studies, not only because spectral resolution must be optimized, but also because it is often necessary to suppress the very large solvent peak. If there are large wings in this peak resulting from problems with B_0 homogeneity, these cannot be suppressed without removing a significant fraction of the whole spectrum at the same time.

Finally, it should be stressed that when designing new probes it is extremely useful to check their performance at all stages of their manufacture using apparatus of the type shown in Fig. 8.5. It should also be pointed out that although the probes described above perform extremely well they are not necessarily optimally designed; no doubt there is scope for further improvement.

Appendix 8.1. Resistance, inductance, and capacitance

Resistance
The reader is probably familiar with the concept of resistance, which is expressed by Ohm's Law

$$V = IR \tag{8.15}$$

where V is the potential difference across a conductor, I is the current flowing through it, and the resistance R of the conductor gives the proportionality between V and I.

At low frequencies the resistance of a conductor is proportional to its length L, and inversely proportional to its cross-section A. Therefore we can write

$$R = \rho L/A \tag{8.16}$$

where ρ is known as the resistivity of the material of which the conductor is made. The conductivity σ of the material is the reciprocal of the resistivity.

Radiofrequency coils are commonly made of copper because of its high conductivity. Silver has a slightly higher conductivity, but copper is generally used because it is less susceptible to paramagnetic impurities.

At high frequencies the skin effect becomes important. This is an effect whereby high-frequency currents flowing along a wire are confined to the outer surface or 'skin' of the wire (see also § 8.5.3). In fact, the current density decreases exponentially with depth of penetration into the wire, and the skin depth δ defines the depth at which the current declines to $1/e$ of its value at the surface. The skin depth is given by

$$\delta = 1/(\pi \sigma \nu \mu)^{1/2} \tag{8.17}$$

where σ is the conductivity of the material, ν is the frequency of the alternating current, and μ is the magnetic permeability. For copper the skin depth is approximately equal to 6.6×10^{-6} m at 100 MHz, and if the diameter of the copper wire through which the alternating current flows is much greater than the skin depth, eqn (8.16) has to be modified; the current flows through a tube of effective cross-section $2\pi r \delta$ where r is the radius of the wire, and therefore

$$R = \rho L/2\pi r \delta. \tag{8.18}$$

In diagrams of electric circuits, resistance is represented by the symbol

Inductance

When a current I flows through a coil a magnetic field is generated in the vicinity of the coil. The energy stored within the field has magnitude $\frac{1}{2}LI^2$, where L is a parameter characteristic of the coil and is known as the 'self-inductance' or simply the inductance. The inductance is also equal to the magnetic flux linkage associated with the coil when unit current flows through it.

For a solenoid of length $2g$, radius a, and n turns, the field B_1 within it is given by

$$B_1 = n\mu_0 I/2(a^2 + g^2)^{1/2}. \tag{8.19}$$

Straightforward calculation shows that its inductance L is

$$L = n^2 \mu_0 A/2(a^2 + g^2)^{1/2} \tag{8.20}$$

where A is the cross-sectional area of the solenoid. The unit of inductance is the henry (H), and radiofrequency coils typically have an inductance of the order of 0.1 μH. In diagrams of electric circuits the symbol for inductance is

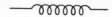

Another important parameter of any coil is its 'quality factor' Q. This is a dimensionless constant which can readily be measured (see § 8.7.1) and is given by the expression

$$Q = \omega L/R \qquad (8.21)$$

where R is the resistance of the coil and ω is the angular frequency of the current passing through it. Now we have seen earlier in this chapter that for optimal signal-to-noise ratios it is essential to maximize the ratio $B_{1(u)} V_s/R^{1/2}$, and since $B_{1(u)}$ is closely related to L it is important to design a coil with a high value of Q. It is also important to appreciate that Q is not a completely reliable guide to the performance of an n.m.r. coil (see Hoult 1978); essentially the purpose of measuring Q should be to obtain a value for the coil resistance.

Capacitance

A capacitor is a device for storing electrical charge, and its capacitance is a measure of the amount of charge that must be placed on it in order to raise its potential by 1 V. The capacitance C can be written

$$C = Q/V \qquad (8.22)$$

and is expressed in farads (F) if the charge Q is in coulombs (C) and the potential V is in volts (V). Capacitors used in probe tuned circuits have typical capacitances of about 10 pF, i.e. 10×10^{-12} F.

A simple type of capacitor consists of two parallel conducting plates, one of which is connected to earth. When a positive charge is placed on the second plate, an equal amount of negative charge flows to the first plate from earth owing to the attraction of the positive charge. The capacitance C can be shown to have the value $C = \epsilon A/d$, where A is the cross-sectional area of the plates, d is the distance between them, and ϵ is the permittivity of the medium between the plates.

The energy stored by a capacitor is equal to $\frac{1}{2} C V^2$. A capacitor is represented diagramatically by the symbol

Therefore the tuned circuit commonly used in n.m.r. probes can be represented as shown in Fig. 8.3, where C_1 represents the tuning capacitance, C_2 the matching capacitance, L the inductance of the radiofrequency coil, and R its resistance.

Further reading

Below are listed a number of texts that the reader may find of value. The more biological texts deal largely with applications of n.m.r. to the structure and function of macromolecules; they contain little, if any, discussion of applications to living systems, because of the relatively recent emergence of these studies. It is also worth noting that Fourier transform n.m.r. was only developed in the late 1960s, and therefore modern n.m.r. techniques are best covered in books written after this time. However, many of the earlier books provide excellent treatments of the theory of n.m.r.

N.m.r. theory and techniques

Abragam, A. (1961). *The principles of nuclear magnetism* Clarendon Press, Oxford. (A standard work for n.m.r. specialists, but rather complicated!).

Becker, E.D. (1969). *High resolution n.m.r.* Academic Press, New York.

Bovey, F.A. (1969). *Nuclear magnetic resonance spectroscopy*. Academic Press, New York.

Carrington, A. and McLachlan, A.D. (1967). *Introduction to magnetic resonance*. Harper and Row, London.

Emsley, J.W., Feeney, J. and Sutcliffe, L.H. (1965). *High resolution nuclear magnetic resonance spectroscopy*. Pergamon Press.

Farrar, T.C. and Becker, E.D. (1971). *Pulse and fourier transform n.m.r.* Academic Press, New York.

Jackman, L.M. and Sternhell, S. (1969). *Applications of nuclear magnetic resonance spectroscopy in organic chemistry*. Academic Press, New York.

Lyndon-Bell, R.M., and Harris, R.K. (1969). *Nuclear magnetic resonance spectroscopy*. Thomas Nelson and Sons, London.

Martin, M.L., Delpuech, J.–J., and Martin, G.J. (1980). *Practical n.m.r. spectroscopy*. Heyden and Son Ltd., London.

Mullen, K., and Pregosin, P.S. (1976). *Fourier transform n.m.r. techniques: a practical approach*. Academic Press, London.

Pople, J.A., Schneider, W.G., and Bernstein, H.J. (1959). *High resolution nuclear magnetic resonance*. McGraw-Hill Book Co., New York, Toronto, and London.

Shaw, D. (1976). *Fourier transform n.m.r. spectroscopy*. Elsevier/North Holland, Amsterdam.

Slichter, C.P. (1963). *Principles of magnetic resonance*. Harper and Row, New York. (2nd edition, 1978, Springer-Verlag Berlin).

N.m.r. and its biological applications

Dwek, R.A. (1973). *Nuclear magnetic resonance in biochemistry*. Oxford University Press, London.

James, T.L. (1975). *Nuclear magnetic resonance in biochemistry: principles and applications*. Academic Press, New York.

Knowles, P.F., Marsh, D., and Rattle, H.W.E. (1976). *Magnetic resonance of biomolecules*. John Wiley, London.

Wuthrich, K. (1976). *N.m.r. in biological research: peptides and proteins*. North-Holland Publishing Co. Amsterdam.

Books presenting n.m.r. reviews

N.m.r. in biology (1977). (eds. R.A. Dwek, I.D. Campbell, R.E. Richards, and R.J.P. Williams). Academic Press, London.

Biological applications of magnetic resonance (1979). (ed. R.G. Shulman). Academic Press, New York.

Magnetic resonance in biology (1980). (ed. J.S. Cohen). John Wiley and Sons, New York.

Reviews of n.m.r. applications to living systems

Burt, C.T., Cohen, S.M., and Bárány, M. (1979). Analysis of intact tissue with ^{31}P n.m.r. *Ann. Rev. Biophys. Bioeng.* **8**, 1.

Gadian, D.G., and Radda, G.K. (1981). N.m.r. studies of tissue metabolism. *Ann. Rev. Biochem.* **50**, 69.

Gadian, D.G., Radda, G.K., Richards, R.E., and Seeley, P.J. (1979). ^{31}P n.m.r. in living tissue – the road from a promising to an important tool in biology. In *Biological applications of magnetic resonance.* (ed. R.G. Shulman). Academic Press, New York. p. 463.

Griffiths, J.R., and Iles, R.A. (1980). Nuclear magnetic resonance – a 'magnetic eye' on metabolism. *Clin. Sci.* **59**, 225.

Hoult, D.I. (1980). Medical Imaging by n.m.r. In *Magnetic resonance in biology* (ed. J.S. Cohen). John Wiley and Sons, New York. p. 70.

Scott, A.I., and Baxter, R.L. (1981). Applications of ^{13}C n.m.r. to metabolic studies. *Ann. Rev. Biophys. Bioeng.* **10**, 151.

Shulman, R.G., Brown, T.R., Ugurbil, K., Ogawa, S., Cohen, S.M., and den Hollander, J.A. (1979). Cellular applications of ^{31}P and ^{13}C nuclear magnetic resonance. *Science* **205**, 160.

Ugurbil, K., Shulman, R.G., and Brown, T.R. (1979). High-resolution ^{31}P and ^{13}C nuclear magnetic resonance studies of *Escherichia coli* cells *in vivo*. In *Biological applications of magnetic resonance.* (ed. R.G. Shulman). Academic Press, New York. p. 537.

Various metabolic studies, together with imaging methods and applications, are reviewed in *N.m.r. of intact biological systems* (eds. R.J.P. Williams, E.R. Andrew, and G.K. Radda). *Phil Trans. R. Soc. Lond. B.* (1980) **289**.

References

Ackerman, J.J.H., Bore, P.J., Gadian, D.G., Grove, T.H., and Radda, G.K. (1980a). *Phil. Trans. R. Soc. Lond. B* **289**, 425.

——, Gadian, D.G., Radda, G.K., and Wong, G.G. (1981). *J. magn. Reson.* **42**, 498.

——, Grove, T.H., Wong, G.G., Gadian, D.G., and Radda, G.K. (1980b). *Nature (Lond.)* **283**, 167.

Akerboom, T.P.M., Van der Meer, R., and Tager, J.M. (1979). *Techniques in metabolic research B* **205**, 1.

Alger, J. R., Sillerud, L.O., Behar, K.L., Gillies, R.J., Shulman, R.G., Gordon, R.E., Shaw, D., and Hanley, P.E. (1981). *Science*. In press.

Bailey, I.A., Williams. S.R., Radda, G.K., and Gadian, D.G. (1981). *Biochem. J.* **196**, 171.

Balaban, R.S., Gadian, D.G., and Radda, G.K. (1981a). *Kidney int.*, in press.

——, ——, ——, and Wong, G.G. (1981b). In press. *Anal. Biochem.*

Bárány, M., Bárány, K., Burt, C.T., Glonek, T., and Myers, T.C. (1975). *J. Supramol. Struct.* **3**, 125.

Battocletti, J.H., Halbach, R.E., Sances, A., Jr., Larson, S.J., Bowman, R.L., and Kudravcev, V. (1979). *Med. Biol. Eng. Comput.* **17**, 183.

Becker, E.D., Ferretti, J.A., and Gambhir, P.N. (1979). *Anal. Chem.* **51**, 1413.

Bendel, P., Lai, C., and Lauterbur, P.C. (1980). *J. magn. Reson.* **38**, 343.

Bertholdi, E. and Ernst, R.R. (1973). *J. magn. Reson.* **11**, 9.

Bloch, F. (1946). *Phys. Rev.* **70**, 460.

——, Hansen, W.W., and Packard, M. (1946). *Phys. Rev.* **70**, 474.

Bore, P.J., Sehr, P.A., Chan, L., Thulborn, K.R., Ross, B.D., and Radda, G.K. (1981). *Transplantation Proc.* **13**, 707.

Bottomley, P.A. and Andrew, E.R. (1978). *Phys. Med. Biol.* **23**, 630.

Brindle, K.M., Brown, F.F., Campbell, I.D., Foxall, D.L. and Simpson, R.J. (1980). *Biochem. Soc. Trans.* **8**, 646.

——, ——, ——, Grathwohl, C., and Kuchel, P.W. (1979). *Biochem. J.* **180**, 37.

Brown, F.F. and Campbell, I.D. (1976). *FEBS Lett.* **65**, 322.

——, and ——, (1980). *Phil. Trans. R. Soc. Lond. B* **289**, 395.

——, ——, Henson, R., Hirst, C.W.J., and Richards, R.E. (1973). *Eur. J. Biochem.* **38**, 54.

——, ——, Kuchel, P.W., and Rabenstein, D.C. (1977a). *FEBS Lett.* **82**, 12.

Brown, T.R. Ugurbil, K., and Shulman, R.G. (1977b). *Proc. Natl. Acad. Sci. USA* **74**, 5551.

Brown and Kushmerick (1981). Unpublished observations.

Brunner, P. and Ernst, R.R. (1979). *J. magn. Reson.* **33**, 83.

Budinger, T. (1979). *IEEE Trans. Nucl. Sci.* **26**, 2821.

Burt, C.T., Glonek, T., and Bárány, M. (1976a). *J. Biol. Chem.* **251**, 2584.

——, ——, and —— (1976b). *Biochemistry* **15**, 4850.

Busby, S.J.W., Gadian, D.G., Radda, G.K., Richards, R.E., and Seeley, P.J. (1978). *Biochem. J.* **170**, 103.

——, and Radda, G.K. (1976). *Curr. Top. cell. Regul.* **10**, 89.

Bystrov, V. (1976). *Progr. nucl. magn. reson. Spectrosc.* **10**, 41.

Campbell, I.D. (1977). In *n.m.r. in biology* (eds. R.A.Dwek, I.D. Campbell, R.E. Richards, and R.J.P. Williams), p. 33. Academic Press, London.

——, and Dobson, C.M. (1979). *Methods biochem. Anal.* **25**, 1.

——, ——, Ratcliffe, R.G., and Williams, R.J.P. (1978). *J. magn. Reson.* **29**, 397.

——, ——, Williams, R.J.P., and Xavier, A.V. (1973). *J. magn. Reson.* **11**, 172.

——, ——, ——, and Wright, P.E. (1975). *FEBS Lett.* **57**, 96.

Carr, H.Y. and Purcell, E.M. (1954). *Phys. Rev.* **94**, 630.

Carter, B.L., Morehead, J., Wolpert, S.M., Hammerschlag, S.B., Griffiths, H.J., and Kahn, P.C. (1977). *Cross-sectional anatomy, computed tomography, and ultrasound correlation*. Appleton-Century-Crofts, New York.

Casey, R.P., Njus, D., Radda, G.K., and Sehr, P.A. (1977). *Biochemistry* **16**, 972.

Chalovich, J.M. and Bárány, M. (1980). *Arch. Biochem. Biophys.* **199**, 615.

——, Burt, C.T., Cohen, S.M., Glonek, T., and Bárány, M. (1977). *Arch. Biochem. Biophys.* **182**, 683.

Chan, L., French, M.E., Gadian, D.G., Morris, P.J., Radda, G.K., Bore, P.J., Ross, B.D., and Styles P. (1981). In *Organ transplantation III* (eds. D.E. Pegg, I. Jacobson, and N.A. Halasz). MTP Press Ltd. In press.

Chance, B. and Williams, G.R. (1956). *Adv. Enzymol.* **17**, 65.

Chiarotti, G., Cristiani, G., and Giuletto, L. (1955). *Nuovo Cim.* **1**, 863.

Cohen, R.D. and Iles, R.A. (1975). *Crit. Rev. clin. lab. Sci.* **6**, 101.

Cohen, S.M. and Burt, C.T. (1977). *Proc. Natl. Acad. Sci. USA* **74**, 4271.

——, Ogawa, S., Rottenberg, H., Glynn, P., Yamane, T., Brown, T.R., Shulman, R.G., and Williamson, J.R. (1978). *Nature (Lond.)* **273**, 554.

——, ——, and Shulman, R. G. (1979a). *Proc. Natl. Acad. Sci. USA* **76**, 1603.

——, Shulman, R.G., and McLaughlin, A.C. (1979b). *Proc. Natl. Acad. Sci. USA* **76**, 4808.

Cohn, M. and Hughes, T.R. (1962). *J. biol. Chem.* **237**, 176.

——, and Rao, B.D.N. (1979). *Bull. magn. Reson.* **1**, 38.

Cooley, J.W. and Tukey, J.W. (1965). *Math. Comput.* **19**, 297.

Cooper, J.W. (1976). In *Topics in carbon-13 n.m.r. spectroscopy* Vol. 2. (ed. G.C. Levy), p. 392. John Wiley and Sons, New York.

Costa, J.L., Dobson, C.M., Kirk, K.L., Poulsen, F.M., Valeri, V.R., and Vecchione, J.J. (1979). *FEBS Lett.* **99**, 141.

Cresshull, I., Dawson, M.J., Edwards, R.H.T., Gadian, D.G., Gordon, R.E., Radda, G.K., Shaw, D., and Wilkie, D.R. (1981). *J. Physiol.* In press.

Dadok, J. and Sprecher, R.F. (1974). *J. magn. Reson.* **13**, 243.

Damadian, R. (1971). *Science* **171**, 1151.

Dawson, M.J., Gadian, D.G., and Wilkie, D.R. (1977a). *J. Physiol.* **267**, 703.

——, ——, and ——. (1977b). In *N.m.r. in biology* (eds. R.A. Dwek, I.D. Campbell, R.E. Richards, and R.J.P. Williams), p. 289. Academic Press, London.

——, ——, and ——, (1978). *Nature (Lond.)* **274**, 861.

——, ——, and ——. (1980a). *J. Physiol.* **299**, 465.

——, ——, and ——, (1980b). *Phil. Trans. R. Soc. Lond.* B **289**, 445.

Delayre, J.L., Ingwall, J.S., Malloy, C., and Fossel, E.T. (1981). *Science.* **212**, 935.

DeMarco, A. and Wuthrich, K. (1976). *J. magn. Reson.* **24**, 201.

Dobson, C.M. (1977). In *N.m.r. in biology* (eds. R.A. Dwek, I.D. Campbell, R.E. Richards, and R.J.P. Williams), p. 63. Academic Press, London.

Dube, G.P., Gadian, D.G., Matthews, P.M., Radda, G.K., Schwartz, A., Seymour, A-M., and Williams, S.R. (1981). In *Non-invasive probes of tissue metabolism* (ed. J.S. Cohen), Wiley Interscience, New York. In press.

Dwek, R.A. (1973). *Nuclear magnetic resonance in biochemistry*. Clarendon Press, Oxford.

Eakin, R.T., Morgan, L.O., Gregg, C.T., and Matwiyoff, N.A. (1972). *FEBS Lett.* **28**, 259.

Edelstein, W.A., Hutchison, J.M.S., Johnson, G., and Redpath, T. (1980). *Phys. Med. Biol.* **25**, 751.

Egan, W., Shindo, H., and Cohen, J.S. (1977). *Ann. Rev. Biophys. Bioeng.* **6**, 383.

Ernst, R.R. (1966). *Adv. mag. Reson.* **2**, 1.

——, and Anderson, W.A. (1966). *Rev. sci. Instrum.* **37**, 93.

Feeney, J. (1975). *Proc. R. Soc. Lond.* A **345**, 61.

Ferrige, A.G. and Lindon, J.C. (1978). *J. magn. Reson.* **31**, 337.

Finch, E.D. (1979). In *The Aqueous Cytoplasm*, Contemporary Biophysics Series, Vol. 1. (ed. A.D. Keith) p. 61. Marcel Dekker Inc. New York.

Forsen, S. and Hoffman, R.A. (1963). *J. chem. Phys.* **39**, 2892.

——, (1964). *J. Chem. Phys.* **40**, 1189.

Fossel, E.T., Morgan, H.E., and Ingwall, J.S. (1980). *Proc. Natl. Acad. Sci. USA* **77**, 3654.

Freeman, R., Kempsall, S.P., and Levitt, M.H. (1980). *J. magn. Reson.* **38**, 453.

Gadian, D.G. (1977). *Contemp. Phys.* **18**, 351.

——, and Radda, G.K. (1981). *Ann. Rev. Biochem.* **50**, 69.

——, ——, Brown, T.R., Chance, E.M., Dawson, M.J., and Wilkie, D.R. (1981). *Biochem J.* **194**, 215.

——, ——, Richards R.E., and Seeley, P.J. (1979). In *Biological applications of magnetic resonance* (ed. R.G. Shulman), p. 463. Academic Press, New York.

——, and Robinson, F.N.H. (1979). *J. magn. Reson.* **34**, 449.

Garlick, P.B. (1979). *D. Phil. Thesis*, Oxford University.

——, Radda, G.K., and Seeley, P.J. (1979). *Biochem J.* **184**, 547.

Gillies, R.J. and Deamer, D.W. (1979). *Curr. Top. Bioenerg.* **9**, 63.

Glonek, T. and van Wezer, J.R. (1974). *J. magn. Reson.* **13**, 390.

Gordon, R.E., Hanley, P., Shaw, D., Gadian, D.G., Radda, G.K. Styles, P., Bore, P.J., and Chan, L. (1980). *Nature (Lond.)* **287**, 367.

Grove, T.H., Ackerman, J.J.H., Radda, G.K., and Bore, P.J. (1980). *Proc. Natl. Acad. Sci. USA* **77**, 299.

Gupta, R.J. and Moore, R.D. (1980). *J. biol. Chem.* **255**, 3987.

Hahn, E.L. (1950). *Phys. Rev.* **80**, 580.

Hawkes, R.C., Holland, G.N., Moore, W.S., and Worthington, B.S. (1980). *J. Comput. Assist. Tomogr.* **4**, 577.

Herzfeld, J., Roufosse, A., Haberkorn, R.A., Griffin, R.G., and Glimcher, M.J. (1980). *Phil. Trans. R. Soc. Lond. B* **289**, 459.

Hinshaw, W.S., Andrew, E.R., Bottomley, P.A., Holland, G.N., Moore, W.S., and Worthington, B.S. (1979). *Br. J. Radiol.* **52**, 36.

——, Bottomley, P.A., and Holland, G.N. (1977). *Nature (Lond.)* **270**, 722.

Holland, G.N., Hawkes, R.C., and Moore, W.S. (1980). *J. comput. assist. Tomogr.* **4**, 429.

Hollis, D.P. (1979). *Bull. magn. Reson.* **1**, 27.

Hollis, D.P., Nunnally, R.L., Taylor, G.J., Weisfeldt, M.L., and Jacobus, W.E. (1978). *J. magn. Reson.* **29**, 319.

Hoult, D.I. (1978). *Progr. n.m.r. Spectrosc.* **12**, 41.

——, (1980a). *Phil. Trans. R. Soc. Lond. B.* **289**, 543.

——, (1980b). In *Magnetic resonance in biology* (ed. J.S. Cohen), p. 70. John Wiley, New York.

——, Busby, S.J.W., Gadian, D.G., Radda, G.K., Richards, R.E., and Seeley, P.J. (1974). *Nature (Lond.)* **252**, 285.

——, and Lauterbur, P.C. (1979). *J. magn. Reson.* **34**, 425.

——, and Richards, R.E. (1975). *Proc. R. Soc. Lond. A* **344**, 311.

——, and ——, (1976). *J. magn. Reson.* **24**, 71.

Iles, R.A., Griffiths, J.R., Stevens, A.N., Gadian, D.G., and Porteous, R. (1980). *Biochem. J.* **192**, 191.

Illingworth, J.A. (1981). *Biochem. J.* **195**, 259.

Jacobson, L., and Cohen, J.S. (1981). *Bioscience Reports* **1**, 141.

Jacobus, W.E., Pores, I.H., Taylor, G.J., Nunnally, R.L., Hollis, D.P., and Weisfeldt, M.L. (1978). *J. mol. cell. Cardiol.* **10** (Suppl. 1), 39.

James, T.L. (1975). *Nuclear magnetic resonance in biochemistry: principles and applications.* Academic Press, New York.

Jobsis, F.F., and Duffield, J.C. (1967). *J. Gen. Physiol.* **50**, 1009.
Jones, D.W. and Child, T.F. (1976). *Adv. magn. Reson.* **8**, 123.
Lauterbur, P.C. (1973). *Nature (Lond.)* **242**, 190.
Lawson, J.W.R., and Veech, R.L. (1979). *J. biol. Chem.* **254**, 6528.
Levy, G.C. and Peat, I.R. (1975). *J. magn. Reson.* **18**, 500.
Ling, C.R., Foster, M.A., and Hutchison, J.M.S. (1980). *Phys. Med. Bull.* **25**, 748.
Luz, Z. and Meiboom, S. (1964). *J. chem. Phys.* **40**, 2686.
McLaughlin, A.C., Takeda, H., and Chance, B. (1979). *Proc. Natl. Acad. Sci. USA* **76**, 5445.
Meakin, P. and Jesson, J.P. (1973). *J. magn. Reson.* **10**, 290.
Meiboom, S. and Gill, D. (1958). *Rev. sci. Instrum.* **29**, 688.
Moon, R.B. and Richards, J.H. (1973). *J. biol. Chem.* **248**, 7276.
Moore, W.J. (1962). *Physical chemistry*, Problem No. 36, Chapter 14. Longmans Green, Harlow, Essex.
Morse, O.C. and Singer, J.R. (1970). *Science* **170**, 440.
Morris, G.H., and Freeman, R. (1979). *J. Am. Chem. Soc.* **101**, 760.
Navon, G., Ogawa, S., Shulman, R.G., and Yamane, T. (1977a). *Proc. Natl. Acad. Sci. USA* **74**, 87.
——, ——, ——, and ——, (1977b). *Proc. Natl. Acad. Sci. USA*. **75**, 1796.
——, ——, ——, and ——, (1977c). *Proc. Natl. Acad. Sci. USA* **74**, 888.
Newsholme, E.A., and Start, C. (1973). *Regulation in Metabolism*. John Wiley and Sons, London.
Njus, D., Sehr, P.A., Radda, G.K., Ritchie, G.A., and Seeley, P.J. (1978). *Biochemistry* **17**, 4337.
Norwood, W.I., Norwood, C.R., Ingwall, J.S., Castaneda, A.R., and Fossel, E.T. (1979). *J. Thorac. Cardiovasc. Surg.* **78**, 823.
Nuccitelli, R., Webb, D.J., Lagier, S.T., and Matson, G.B. (1981). *Proc. Natl. Acad. Sci. USA*. In press.
Nunnally, R.L. and Hollis, D.P. (1979). *Biochemistry* **18**, 3642.
Odeblad, E. and Lindstrom, G. (1955). *Acta Radiol.* **43**, 469.
Ogawa, S., Rottenberg, H., Brown, T.R., Shulman, R.G., Castillo, C.L., and Glynn, P. (1978). *Proc. Natl. Acad. Sci. USA* **75**, 1796.
Ogino, T., Arata, Y., and Fujiwara, S. (1980). *Biochemistry* **19**, 3684.
Oldfield, E. and Meadows, M. (1978). *J. magn. Reson.* **31**, 327.
Pollard, H.B., Shindo, H., Creutz, C.E., Pazoles, C.J., and Cohen, J.S. (1979). *J. Biol. Chem.* **254**, 1170.
Pople, J.A., Schneider, W.G., and Bernstein, H.J. (1959). *High resolution nuclear magnetic resonance*. McGraw-Hill, New York.
Purcell, E.M., Torrey, H.C., and Pound, R.V. (1946). *Phys. Rev.* **69**, 37.
Rabenstein, D.L., and Isab, A.A. (1979). *J. magn. Reson.* **36**, 281.
——, and Nakashima, T.T. (1979). *Anal. Chem.* **51**, 1465A.
Radda, G.K., Ackerman, J.J.H., Bore, P.J., Sehr, P.A., Wong, G.G., Ross, B.D., Green, Y., Bartlett, S., and Lowry, M. (1980). *Int. J. Biochem.* **12**, 277.
Redfield, A.G., Kunz, S.D., and Ralph, E.K. (1975). *J. magn. Reson.* **19**, 114.
Roos, A. and Boron, W.F. (1981). *Physiol. Rev.* **61**, 296.
Ross, B.D. (1972). *Perfusion techniques in biochemistry*. Clarendon Press, Oxford.
Ross, B.D., Radda, G.K., Gadian, D.G., Rocker, G., Esiri, M., and Falconer-Smith, J. (1981). *New Eng. J. Med.* **304**, 1338.
Saunders, M., Wishnia, A., and Kirkwood, J.G. (1957). *J. Am. Chem. Soc.* **79**, 3289.
Schaefer, J., Stejskal, E.O., and McKay (1979). *Biochem. Biophys. Res. Commun.* **88**, 274.
Seeley, P.J., Busby, S.J.W., Gadian, D.G., Radda, G.K., and Richards, R.E. (1976). *Biochem. Soc. Trans.* **4**, 62.

Sequin, U. and Scott, A.I. (1974). *Science* **186**, 101.

Shaw, D. (1976). *Fourier transform n.m.r. spectroscopy*. Elsevier–North-Holland, Amsterdam.

Shaw, T.M. and Elsken, R.H. (1950). *J. chem. Phys.* **18**, 1113.

Shporer, M. and Civan, M.M. (1977). *Curr. Top. Membr. Transp.* **9**, 1.

Shulman, R.G., Brown, T.R., Ugurbil, K., Ogawa, S., Cohen, S.M., and den Hollander, J.A. (1979). *Science* **205**, 160.

Sies, H., and Brauser, B. (1980). *Methods biochem. Anal.* **26**, 285.

Singer, J.R. (1959). *Science* **130**, 1652.

Slichter, C.R. (1963). *Principles of magnetic resonance*. Harper and Row, New York. (2nd edition, 1978 Springer-Verlag *Berlin*).

Smith, F.W., Hutchison, J.M.S., Mallard, J.R., Johnson, G., Redpath, T.W., Selbie, R.D., Reid, A., and Smith, C.C. (1981a). *Br. Med. J.* **282**, 510.

Smith, F.W., Hutchison, J.M.S., Johnson, G., Mallard, J.R., Reid, A., Redpath, T.W., and Selbie R.D. (1981*b*). *Lancet* i, 78.

Smith, F.W., Mallard, J.R., Reid, A., and Hutchison, J.M.S. (1981*c*). *Lancet* i 963.

Stejskal, E.O. and Schaefer, J. (1974). *J. magn. Reson.* **14**, 160.

Styles, P., Grathwohl, C., and Brown, F.F. (1979). *J. magn. Reson.* **35**, 329.

Swift, T.J. and Connick, R.E. (1962). *J. chem. Phys.* **33**, 303.

Thulborn, K.R., du Boulay, G.H., and Radda, G.K. (1981a). *Proceedings of the Xth Int. Symp. on Cerebral Blood Flow and Metabolism*. In press.

Thulborn, K.R., Waterton, J.C., Styles, P., and Radda, G.K. (1981b). *Biochem. Soc. Trans.* **9**, 233.

Ugurbil, K., Brown, T.R., den Hollander, J.A., Glynn, P., and Shulman, R.G. (1978a). *Proc. Natl. Acad. Sci. USA* **75**, 3742.

——, Holmsen, H., and Shulman, R.G. (1979a). *Proc. Natl. Acad. Sci. USA* **76**, 2227.

——, Rottenberg, H., Glynn, P., and Shulman, R.G. (1978b). *Proc. Natl. Acad. Sci. USA* **75**, 2244.

——, Shulman, R.G., and Brown, T.R. (1979b). In *Biological applications of magnetic resonance* (ed. R.G. Shulman), p. 537. Academic Press, New York.

Veech, R.L., Lawson, J.W.R., Cornell, N.W., and Krebs, H.A. (1979). *J. biol. Chem.* **254**, 6538.

Wuthrich, K. (1976). *N.m.r. in biological research: peptides and proteins*. North-Holland, Amsterdam.

——, Wagner, G., Richarz, R., and de Marco, A. (1977). In *N.m.r. in biology*. (eds. R.A. Dwek, I.D. Campbell, R.E. Richards, and R.J.P. Williams). p. 51, Academic Press, London.

Index